Essentials of
CHEMICAL BONDING

Essentials of
CHEMICAL BONDING

John Stanton

Kruger Brentt
Publishers
2025

Kruger Brentt Publishers UK. LTD.
Company Number 9728962

Regd. Office: 68 St Margarets Road, Edgware, Middlesex HA8 9UU

© 2025 AUTHOR

ISBN: 978-1-78715-293-9

For information on all our publications visit our website at http://krugerbrentt.com/

PREFACE

"Essentials of Chemical Bonding" is an exploration into the fundamental principles that govern the formation, structure, and properties of molecules, offering readers a comprehensive understanding of the intricate interactions between atoms and molecules. This book serves as an essential resource for students, educators, and researchers in chemistry and related fields, providing a foundational framework for understanding the behavior of matter at the molecular level.

Chemical bonding lies at the heart of chemistry, shaping the properties and reactivity of substances in the natural world and in the laboratory. From the simplest diatomic molecules to the complex structures of biological macromolecules, the nature of chemical bonds determines the stability, solubility, and functionality of substances across diverse applications and disciplines.

As editors of this volume, we have endeavored to compile a diverse array of perspectives, methodologies, and case studies from leading experts and practitioners in the field of chemical bonding. Through a blend of theoretical concepts, experimental techniques, and practical applications, this book covers a wide range of topics, including covalent bonding, ionic bonding, metallic bonding, molecular geometry, and intermolecular forces.

Our overarching goal in presenting this book is to provide readers with a comprehensive and accessible resource that fosters a deeper understanding of chemical bonding and its role in shaping the properties and behavior of matter. By demystifying complex concepts, providing real-world examples, and offering practical insights, we aim to empower students, educators, and researchers to navigate the intricacies of chemical bonding with confidence and clarity.

We extend our sincere gratitude to all the contributors who have generously shared their expertise, experiences, and insights in the field of chemical bonding. It is our hope that this book will serve as a valuable reference and guide for learners and practitioners alike, inspiring curiosity, critical thinking, and innovation in the study of chemical bonding and its myriad applications.

John Stanton

CONTENTS

01 | INTRODUCTION TO CHEMICAL BONDING

1.1 DEFINITION AND SIGNIFICANCE OF CHEMICAL BONDING

1.1.1 Introduction

Chemical bonding refers to the formation of a chemical bond between two or more atoms, molecules or ions to give rise to a chemical compound. These chemical bonds are what keep the atoms together in the resulting compound. Chemical Bonding as the name suggests means the interaction of different elements or compounds which defines the properties of matter. Chemical bonds are formed when either at least one electron is lost to another atom, obtaining at least one electron from a different atom, or transferring one electron to another atom. In this article, we will learn about the theories of chemical bonding such as Lewis's theory which explains the Lewis structure of any compound, Kossel's Theory, and Fajan's rule. So, let's begin the journey of Chemical Bonding.

1.1.2 Chemical Bonding

Chemical bonding is the formation of a chemical bond between two or more atoms, molecules, or ions that results in the formation of a chemical compound. These chemical bonds are what hold the atoms in the resulting compound together.

Chemical bonding is the attractive force that holds various constituents (atoms, ions, etc.) together and stabilizes them through the overall loss of energy. As a result, chemical compounds are dependent on the strength of the chemical bonds between their constituents; the stronger the bonding between the constituents, the more stable the resulting compound.

The inverse is also true: if the chemical bonding between the constituents is weak, the resulting compound will be unstable and will easily undergo another reaction to produce a more stable chemical compound (containing stronger bonds). Atoms try to lose energy in order to find stability. When one form of matter interacts with another, a force is exerted on the first. When the forces of nature are attractive, the energy decreases. When the forces of nature are repulsive, the energy increases. The chemical bond is the attractive force that holds two atoms together.

A common atom contains a nucleus composed of protons and neutrons, with electrons in certain energy levels revolving around the nucleus. In this section, the main focus will be on these electrons. Elements are distinguishable from each other due to their "electron cloud," or the area where electrons move around the nucleus of an atom. Because each element has a distinct electron cloud, this determines their chemical properties as well as the extent of their reactivity (i.e. noble gases are inert/not reactive while alkaline metals are highly reactive). In chemical bonding, only valence electrons, electrons located in the orbitals of the outermost energy level (valence shell) of an element, are involved.

Lewis Diagrams

Lewis diagrams are graphical representations of elements and their valence electrons. Valance electrons are the electrons that form the outermost shell of an atom. In a Lewis diagram of an element, the symbol of the element is written in the center and the valence electrons are drawn around it as dots. The position of the valence electrons drawn is unimportant. However, the general convention is to start from 12o'clock position and go clockwise direction to 3 o'clock, 6 o'clock, 9 o'clock, and back to 12 o'clock positions respectively. Generally the Roman numeral of the group corresponds with the number of valance electrons of the element.

Below is the periodic table representation of the number of valance electrons. The alkali metals of Group IA have one valance electron, the alkaline-earth metals of Group IIA have 2 valance electrons, Group IIIA has 3 valance electrons, and so on.

Valence Electrons in Each Group

1																	2
1	2											3	4	5	6	7	8
1	2											3	4	5	6	7	8
1	2											3	4	5	6	7	8
1	2											3	4	5	6	7	8
1	2											3	4	5	6	7	8
1	2											3	4	5	6		

Lewis diagrams for Molecular Compounds/Ions

To draw the lewis diagrams for molecular compounds or ions, follow these steps below (we will be using H2O as an example to follow):

1) Count the number of valance electrons of the molecular compound or ion. Remember, if there are two or more of the same elements, then you have to double or multiply by however many atoms there are of the number of valance electrons. Follow the roman numeral group number to see the corresponding number of valance electrons there are for that element.

Valance electrons:

Oxygen (O)--Group VIA: therefore, there are 6 valance electrons

Hydrogen (H)--Group IA: therefore, there is 1 valance electron

NOTE: There are TWO hydrogen atoms, so multiply 1 valance electron X 2 atoms

Total: 6 + 2 = 8 valance electrons

2) If the molecule in question is an ion, remember to add or subract the respective number of electrons to the total from step 1.

For ions, if the ion has a negative charge (anion), add the corresponding number of electrons to the total number of electrons (i.e. if NO_3^- has a negative charge of 1-, then you add 1 extra electron to the total; 5 + 3(6)= 23 +1 = 24 total electrons). A - sign mean the molecule has an overall negative charge, so it must have this extra electron. This is because anions have a higher electron affinity (tendency to gain electrons). Most anions are composed of nonmetals, which have high electronegativity.

If the ion has a positive charge (cation), subtract the corresponding number of electrons to the total number of electrons (i.e. H_3O^+ has a positive charge of 1^+, so you subtract 1 extra electron to the total; 6 + 1(3) = 9 - 1 = 8 total electrons). A $^+$ sign means the molecule has an overall positive charge, so it must be missing one electron. Cations are positive and have weaker electron affinity. They are mostly composed of metals; their atomic radii are larger than the nonmetals. This consequently means that shielding is increased, and electrons have less tendency to be attracted to the "shielded" nucleus.

From our example, water is a neutral molecule, therefore no electrons need to be added or subtracted from the total.

3) Write out the symbols of the elements, making sure all atoms are accounted for (i.e. H_2O, write out O and 2 H's on either side of the oxygen). Start by adding single bonds (1 pair of electrons) to all possible atoms while making sure they follow the octet rule (with the exceptions of the duet rule and other elements mentioned above).

4) If there are any leftover electrons, then add them to the central atom of the molecule (i.e. XeF_4 has 4 extra electrons after being distributed, so the 4 extra electrons are given to Xe: like so. Finally, rearrange the electron pairs into double or triple bonds if possible.

Hydrogen Oxide - Water, H$_2$O

$$H^{\bullet} \; + \; H^{\bullet} \; + \; {\overset{\bullet\bullet}{\underset{\bullet\bullet}{\bullet O \bullet}}}$$

$$H^{\bullet}\bullet{\overset{\bullet\bullet}{\underset{\bullet\bullet}{O}}}\bullet{\bullet}H$$

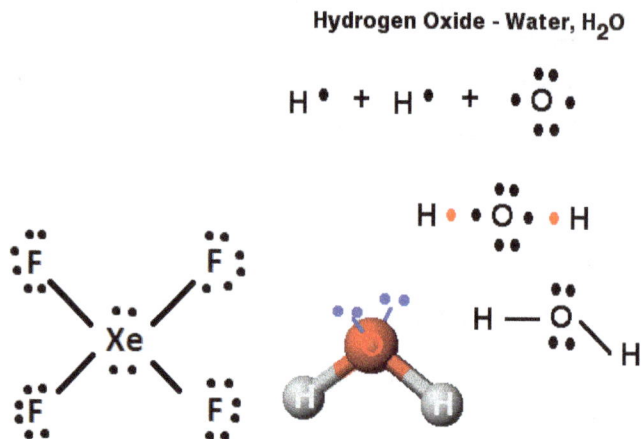

C. Ophardt, c. 2003

Octet Rule

Most elements follow the octet rule in chemical bonding, which means that an element should have contact to eight valence electrons in a bond or exactly fill up its valence shell. Having eight electrons total ensures that the atom is stable. This is the reason why noble gases, a valence electron shell of 8 electrons, are chemically inert; they are already stable and tend to not need the transfer of electrons when bonding with another atom in order to be stable. On the other hand, alkali metals have a valance electron shell of one electron. Since they want to complete the octet rule they often simply lose one electron. This makes them quite reactive because they can easily donate this electron to other elements. This explains the highly reactive properties of the Group IA elements.

Some elements that are exceptions to the octet rule include Aluminum (Al), Phosphorus(P), Sulfur(S), and Xenon (Xe).

Hydrogen(H) and Helium (He) follow the duet rule since their valence shell only allows two electrons. There are no exceptions to the duet rule; hydrogen and helium will always hold a maximum of two electrons.

Ionic Bonding

Ionic bonding is the process of not sharing electrons between two atoms. It occurs between a nonmetal and a metal. Ionic bonding is also known as the process in which electrons are "transferred" to one another because the two atoms have different levels of electron affinity. In the picture below, a sodium (Na) ion and a chlorine (Cl) ion are being combined through ionic bonding. Na+ has less electronegativity due to a large atomic radius and essentially does not want the electron it has. This will easily allow the more electronegative chlorine atom to gain the electron to complete its 3rd energy level. Throughout this process, the transfer of the electron releases energy to the atmosphere.

Ionic Bonding

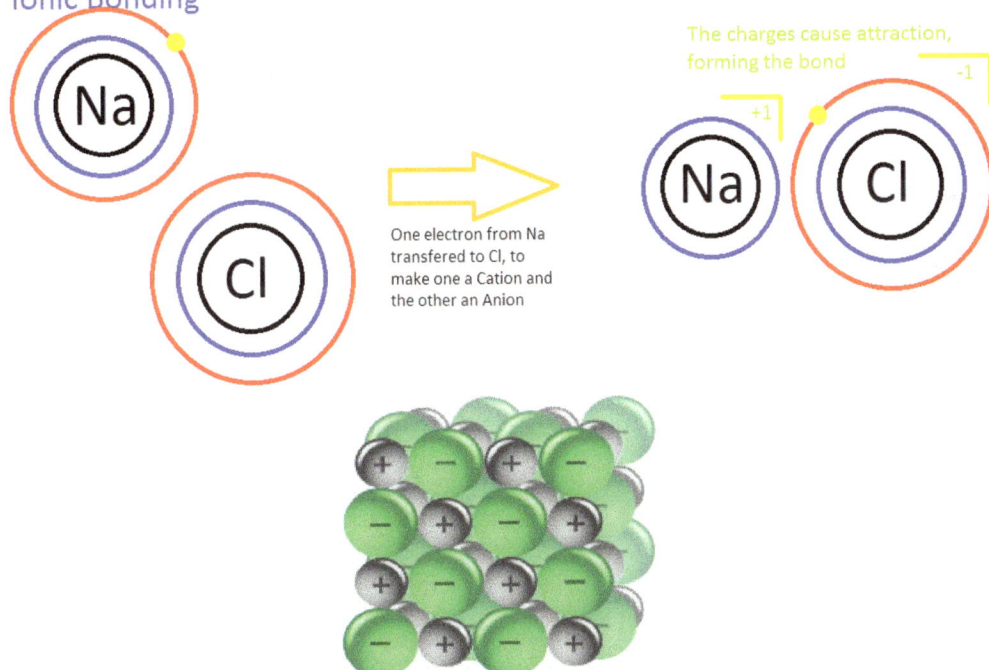

One electron from Na transfered to Cl, to make one a Cation and the other an Anion

The charges cause attraction, forming the bond

Another example of ionic bonding is the crystal lattice structure shown above. The ions are arranged in such a way that shows unifomity and stablity; a physical characteristic in crystals and solids. Moreover, in a concept called "the sea of electrons," it is seen that the molecular structure of metals is composed of stabilized positive ions (cations) and "free-flowing" electrons that weave in-between the cations. This attributes to the metal property of conductivity; the flowing electrons allow the electric current to pass through them. In addition, this explains why strong electrolytes are good conductors. Ionic bonds are easily broken by water because the polarity of the water molecules shield the anions from attracting the cations. Therefore, the ionic compounds dissociate easily in water, and the metallic properties of the compound allow conductivity of the solution.

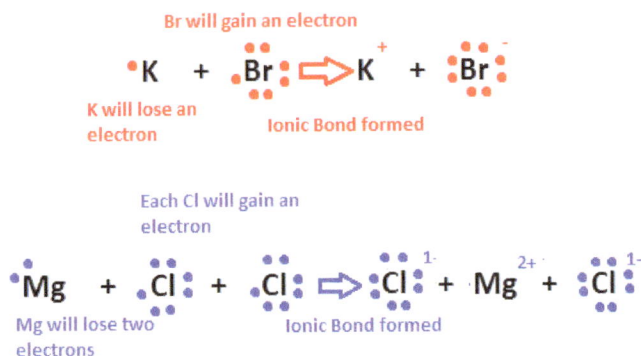

Br will gain an electron

K will lose an electron

Ionic Bond formed

Each Cl will gain an electron

Mg will lose two electrons

Ionic Bond formed

Covalent Bonding

Covalent bonding is the process of sharing of electrons between two atoms. The bonds are typically between a nonmetal and a nonmetal. Since their electronegativities are all within the high range, the electrons are attracted and pulled by both atom's nuceli. In the case of two identical atoms that are bonded to each other (also known as a nonpolar bond, explained later below), they both emit the same force of pull on the electrons, thus there is equal attraction between the two atoms (i.e. oxygen gas, or O_2, have an equal distribution of electron affinity. This makes covalent bonds harder to break.

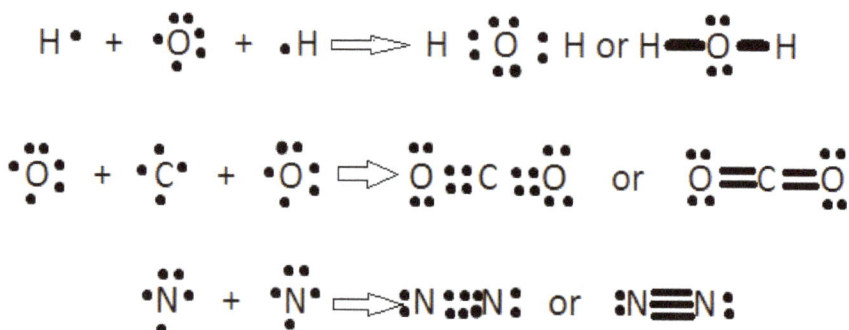

All these diagrams show the formation of covalent bonds as electrons from elements are shared to form single, double, and triple bonds

There are three types of covalent bonds: single, double, and triple bonds. A single bond is composed of 2 bonded electrons. Naturally, a double bond has 4 electrons, and a triple bond has 6 bonded electrons. Because a triple bond will have more strength in electron affinity than a single bond, the attraction to the positively charged nucleus is increased, meaning that the distance from the nucleus to the electrons is less. Simply put, the more bonds or the greater the bond strength, the shorter the bond length will be. In other words:

Bond length: triple bond < double bond < single bond

Polar Covalent Bonding

Polar covalent bonding is the process of unequal sharing of electrons. It is considered the middle ground between ionic bonding and covalent bonding. It happens due to the differing electronegativity values of the two atoms. Because of this, the more electronegative atom will attract and have a stronger pulling force on the electrons. Thus, the electrons will spend more time around this atom.

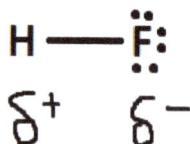

The symbols above indicate that on the flourine side it is slightly negative and the hydrogen side is slightly positive.

Polar and non-polar molecules

Polarity is the competing forces between two atoms for the electrons. It is also known as the polar covalent bond. A molecule is polar when the electrons are attracted to a more electronegative atom due to its greater electron affinity. A nonpolar molecule is a bond between two identical atoms. They are the ideal example of a covalent bond. Some examples are nitrogen gas (N_2), oxygen gas (O_2), and hydrogen gas (H_2).

H 2.1																	He
Li 1.0	Be 1.5											B 2.0	C 2.5	N 3.0	O 3.5	F 4.0	Ne
Na 0.9	Mg 1.2											Al 1.5	Si 1.8	P 2.1	S 2.5	Cl 3.0	Ar
K 0.8	Ca 1.0	Sc 1.3	Ti 1.5	V 1.6	Cr 1.6	Mn 1.5	Fe 1.8	Co 1.9	Ni 1.8	Cu 1.9	Zn 1.6	Ga 1.6	Ge 1.8	As 2.0	Se 2.4	Br 2.8	Kr
Rb 0.8	Sr 1.0	Y 1.2	Zr 1.4	Nb 1.6	Mo 1.8	Tc 1.9	Ru 2.2	Rh 2.2	Pd 2.2	Ag 1.9	Cd 1.7	In 1.7	Sn 1.8	Sb 1.9	Te 2.1	I 2.5	Xe
Cs 0.7	Ba 0.9	Lu	Hf 1.3	Ta 1.5	W 1.7	Re 1.9	Os 2.2	Ir 2.2	Pt 2.2	Au 2.4	Hg 1.9	Tl 1.8	Pb 1.9	Bi 1.9	Po 2.0	At 2.2	Rn
Fr 0.7	Ra 0.9	Lr	Rf	Db	Sg	Bh	Hs	Mt	Ds	Uuu	Uub	Uut	Uuq	Uup	Uuh	Uus	Uuo

One way to figure out what type of bond a molecule has is by determining the difference of the electronegativity values of the molecules.

If the difference is between 0.0-0.3, then the molecule has a non-polar bond.

If the difference is between 0.3-1.7, then the molecule has a polar bond.

If the difference is 1.7 or more, then the molecule has an ionic bond.

1.1.3 Significance of Chemical Bonding

Chemical bonding is a fundamental concept in chemistry, crucial for understanding the structure, properties, and behaviour of matter at the molecular level. It describes the attractive forces that hold atoms together to form compounds, molecules, and solids. The significance of chemical bonding lies in its profound impact on various aspects of chemistry, materials science, and everyday life.

One of the most significant aspects of chemical bonding is its role in determining the stability and reactivity of substances. Atoms combine with each other through chemical bonds to achieve a more stable electronic configuration, typically by attaining a full outer electron shell. This stability is achieved through the sharing, transfer, or redistribution of electrons between atoms. The type of chemical bond formed between atoms—whether it's ionic, covalent, metallic, or hydrogen— dictates the properties of the resulting compounds.

Ionic bonding involves the transfer of electrons from one atom to another, resulting in the formation of positively and negatively charged ions that attract each other through electrostatic forces. This type of bonding is commonly observed in compounds composed of metals and nonmetals and is responsible for the formation of many salts and minerals.

Covalent bonding, on the other hand, involves the sharing of electron pairs between atoms, leading to the formation of molecules. In covalent bonds, atoms share electrons to achieve a more stable configuration, resulting in the formation of discrete molecular entities with unique properties. Covalent bonds are prevalent in organic compounds and many other substances found in nature and industry.

Metallic bonding occurs in metals and alloys, where positively charged metal ions are surrounded by a sea of delocalized electrons. These electrons are free to move throughout the metal lattice, giving rise to properties such as conductivity, malleability, and ductility. Metallic bonding is responsible for the characteristic properties of metals and their widespread use in various applications, from electrical wiring to structural materials.

Hydrogen bonding is a special type of intermolecular force that occurs when a hydrogen atom bonded to a highly electronegative atom (such as oxygen, nitrogen, or fluorine) interacts with another electronegative atom nearby. Hydrogen bonds are responsible for many unique properties of water, including its high boiling point, surface tension, and ability to dissolve a wide range of substances. They also play crucial roles in biological systems, such as protein folding and DNA structure.

Understanding chemical bonding is essential for predicting the properties and behavior of substances, designing new materials, and developing novel technologies. It provides the foundation for many areas of chemistry, including organic chemistry, inorganic chemistry, physical chemistry, and materials science. By elucidating the nature of chemical bonds, scientists and engineers can manipulate molecular structures to create materials with tailored properties for specific applications, ranging from pharmaceuticals and polymers to electronics and nanotechnology.

Chemical bonding is a cornerstone of chemistry, enabling the formation of diverse substances and governing their properties and interactions. Its significance extends far beyond the confines of the laboratory, shaping our understanding of the natural world and driving innovation in fields ranging from materials science to biotechnology. As we continue to explore and harness the power of chemical bonding, we unlock new possibilities for scientific discovery and technological advancement, paving the way for a brighter and more sustainable future.Top of Form

1.2 HISTORICAL OVERVIEW OF THEORIES ON CHEMICAL BONDING

When delving into the history of chemistry, it becomes inherently risky to pinpoint the exact origin of an idea, given that the scientific process inherently thrives on the gradual evolution and refinement of preceding concepts. Nevertheless, certain pivotal moments stand out, shaping our understanding of chemical bonding. In the early 18th century, a renowned publication encapsulates one such moment in the annals of chemistry's development.

In his 1704 publication "Opticks," Sir Isaac Newton mentions a force that foreshadows the modern idea of the chemical bond. In Query 31 of the book, Newton describes forces, beyond magnetism and gravity, that enable particles to interact. In 1718, while translating "Opticks" into French, chemist Étienne François Geoffroy created an Affinity Table. This table provided an initial understanding of the likelihood of certain interactions by tabulating the relative affinity of various substances for each other, thus describing the strength of their interactions.

Although Newton and Geoffroy's work preceded our modern comprehension of elements and compounds, it offered valuable insights into chemical interactions. However, it took over 100 years before the concept of the combining power of elements was understood more comprehensively. In his 1852 paper in the journal "Philosophical Transactions" titled "On a new series of organic bodies containing metals," Edward Frankland described the "combining power of elements," which is now known as valency in chemistry.

Frankland proposed a law suggesting that regardless of the characters of the uniting atoms, the combining power of the attracting element, or valency, is always satisfied by the same number of these atoms. His work implied that each element combines with only a limited number of atoms of another element, hinting at the concept of bonding. However, it was two other scientists who conducted the most significant contemporary research on bonding.

In 1916, the renowned American scientist Gilbert N. Lewis authored a seminal paper titled "The Atom and the Molecule," which delved into fundamental principles of bonding. In this publication, Lewis introduced pivotal concepts that continue to serve as foundational models for understanding electron arrangement at the atomic level. A central tenet of his theory revolved around the valence electrons of atoms, proposing that chemical bonds form through the sharing of electron pairs between atoms. This concept, later coined as a covalent bond by Irving Langmuir, was illustrated by Lewis through his iconic "Lewis dot diagrams," where each shared pair of electrons constituting a covalent bond was represented by a pair of dots.

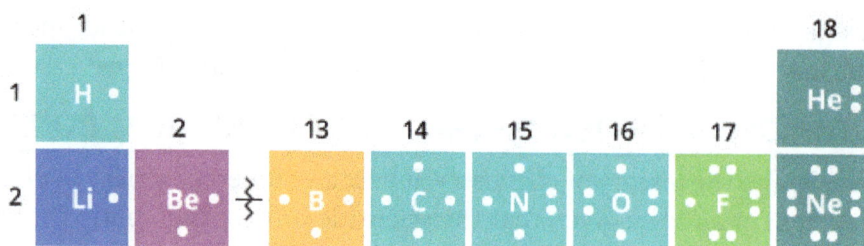

Figure : Lewis dot structures for the elements in the first two periods of the periodic table. The structures are written as the element symbol surrounded by dots that represent the valence electrons.

Lewis also advocated for the concept of 'octets' (sets of eight), emphasizing the significance of a filled valence shell in comprehending electronic configurations and the bonding patterns between atoms. While the notion of the octet had been previously broached by chemists like John Newlands, who recognized its importance, Lewis notably refined and advanced the theory.

1.3 TYPES OF CHEMICAL BONDS

When substances participate in chemical bonding and yield compounds, the stability of the resulting compound can be gauged by the type of chemical bonds it contains.

The type of chemical bonds formed varies in strength and properties. There are 4 primary types of chemical bonds which are formed by atoms or molecules to yield compounds. These types of chemical bonds include

- ◉ Ionic Bonds
- ◉ Covalent Bonds
- ◉ Hydrogen Bonds
- ◉ Polar Bonds

These types of bonds in chemical bonding are formed from the loss, gain or sharing of electrons between two atoms/molecules.

1.3.1 Ionic Bonding

Ionic bonding is a type of chemical bonding which involves a transfer of electrons from one atom or molecule to another. Here, an atom loses an electron, which is, in turn, gained by another atom. When such an electron transfer takes place, one of the atoms develops a negative charge and is now called the anion.

The other atom develops a positive charge and is called the cation. The ionic bond gains strength from the difference in charge between the two atoms, i.e., the greater the charge disparity between the cation and the anion, the stronger the ionic bond.

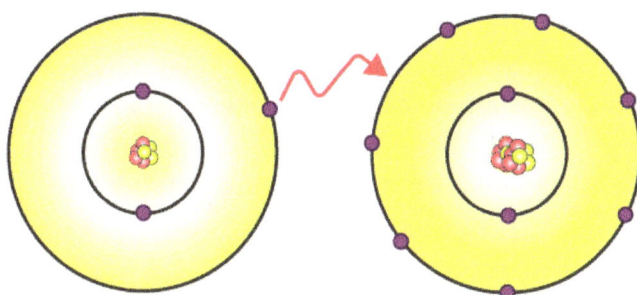

Types of Chemical Bonds – Ionic bonding

1.3.2 Covalent Bonding

A covalent bond indicates the sharing of electrons between atoms. Compounds that contain carbon (also called organic compounds) commonly exhibit this type of chemical bonding. The pair of electrons which are shared by the two atoms now extend around the nuclei of atoms, leading to the creation of a molecule.

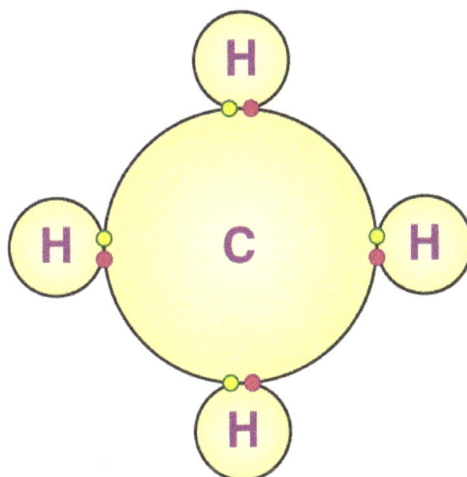

○ Electron from Hydrogen | ● Electron from Carbon

Covalent Bonding

1.3.3 Polar Covalent Bonding

Covalent bonds can be either polar or non-polar in nature. In polar covalent chemical bonding, electrons are shared unequally since the more electronegative atom pulls the electron pair closer to itself and away from the less electronegative atom. Water is an example of such a polar molecule.

A difference in charge arises in different areas of the atom due to the uneven spacing of the electrons between the atoms. One end of the molecule tends to be partially positively charged, and the other end tends to be partially negatively charged.

1.3.4 Hydrogen Bonding

Compared to ionic and covalent bonding, Hydrogen bonding is a weaker form of chemical bonding. It is a type of polar covalent bonding between oxygen and hydrogen, wherein the hydrogen develops a partial positive charge. This implies that the electrons are pulled closer to the more electronegative oxygen atom.

This creates a tendency for the hydrogen to be attracted towards the negative charges of any neighboring atom. This type of chemical bonding is called a hydrogen bond and is responsible for many of the properties exhibited by water.

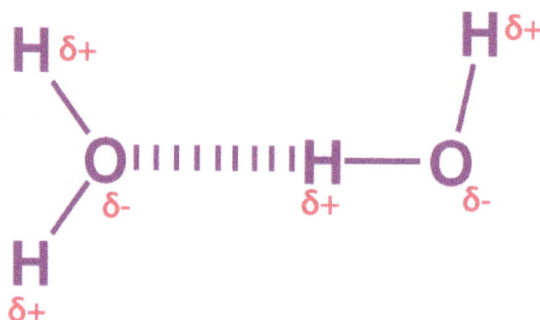

Hydrogen Bonding

1.3.5 Ionic Bond

The bond formed as a result of strong electrostatic forces of attraction between a positively and negatively charged species is called an electrovalent or ionic bond. The positively and negatively charged ions are aggregated in an ordered arrangement called the crystal lattice, which is stabilised by the energy called the Lattice enthalpy.

Conditions for the Formation of an Ionic Bond

- The low ionisation energy of the atom forming the cation.
- High electron gain enthalpy of the atom forming the anion.
- High negative lattice enthalpy of the crystal formed.
- Generally, the ionic bond is formed between a metal cation and a non-metal anion.

1.4 THE ROLE OF CHEMICAL BONDING IN MOLECULAR STRUCTURE

1.4.1 An Overview

- An atom (or) group of atoms having independent existence in nature is defined as molecule.
- Force of attraction that holds two atoms (or) oppositely charged ions in molecule is chemical bond.

- An atom must have eight electrons is its outermost energy level for its stability is called as octet rule.

- The number of electrons an atom loses (or) gains in an ionic bond formation is called electro valency

- $LiCl$ is more covalent in nature and so dissolved in non-polar solvents like alcohol (or) ether.

- Ions which have 18 electrons [$ns^2np^6nd^{10}$] in their outermost main energy level are stable and it is called as pseudo inert gas configuration (or) nickel group configuration.

- The energy liberated when an ionic crystal is formed from an assembly of isolated gaseous ions is called lattice energy

- Lattice energy

$$(\cup) = \left[-No\frac{AZ^+.Z^-e^2}{r} + No\frac{Be^2}{r^n} \right]$$

- The no. of electrons an atom of the element losses (or) gains is known as electrovalency of that element.

- Favourable conditions for cation formation.

  ➤ Low I.P

  ➤ Low charge on ion

  ➤ Big atomic size

  ➤ Inert gas configuration

- Favourable conditions for cation formation.

  ➤ High E.N & E.A

  ➤ Small atomic size

  ➤ Low charge on ion

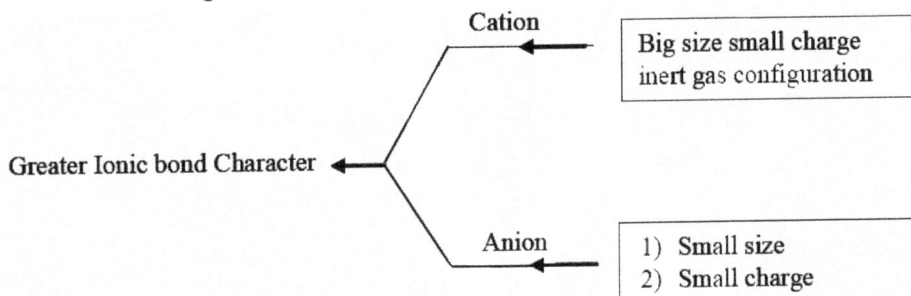

$$\cup = \frac{AZ^+Z^-e^2No}{r_0}\left[1-\frac{1}{n}\right]$$

(Born - Lande equation)

Coordination number of Na+ and Cl^- in NaCl crystal is (6:6)

Coordination number of Cs+ and Cl^- in NaCl crystal is (8:8)

⊙ The total amount of energy change in a reaction remains the same whether the reaction take place in one (or) several stages is known as Hess law of constant heat summation.

⊙ The maximum number of nearest oppositely charged ions surrounding any particular ion in an ionic.

⊙ Crystal is called the co-ordination number of that ion.

⊙ Crystal structure of NaCl is face centered cubic lattice structre.

⊙ Crystal structure of CsCl is body centered cubic lattice structure.

1.4.2 Properties of Ionic Compounds

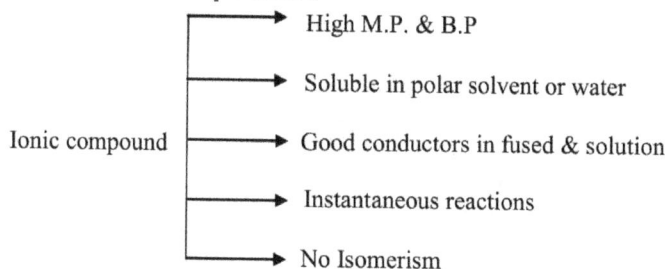

Ionic compound
- High M.P. & B.P
- Soluble in polar solvent or water
- Good conductors in fused & solution
- Instantaneous reactions
- No Isomerism

• Properties of covalent compounds

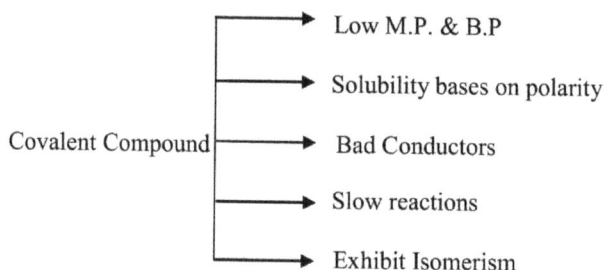

Covalent Compound
- Low M.P. & B.P
- Solubility bases on polarity
- Bad Conductors
- Slow reactions
- Exhibit Isomerism

⊙ The number of electrons an atom of an element contributes in the formation of covalent compound is called covalency of the element.

⊙ The charge possible that an atom in a molecule would have if all the atoms had the same electronegativity is called formal charge.

Formal charge

$$\left(Q_f\right)=\left[N_A - N_M\right]$$

$$=\left[N_A - N_{L.P.} -\frac{1}{2}N_{B.P.}\right]$$

Where

NA = No.

of electrons in the valence shell in free atom NM = No.

of electrons belonging to the atom in molecule NL.P. = No.

of electrons in unshared pairs is lone pairs. NB.P. = No.

of electrons in bond pairs.

◉ The molecule with a smallest formal charges on the atoms, is considered as lower energy structure of a molecule

V.S.E.P.R. theory

Number of bond pairs	Number of lone pairs	Formula	Molecular Shape	Hybridization	Bond Angle	Example
2	0	AB_2	Linear	Sp	180^0	BeCl2, BeF2 ,C2 H2
3	0	AB_3	Plane Triangle	sp^2	120^0	BF3, BCl3
4	0	AB_4	Tetrahedron	sp^3	$109^0, 28^1$	CH_4, NH^+_4, CCl_4
3	1	AB_3E	Trigonal pyramid	sp^3	$107^0, 102^0, 30^1$	$NH_3, H\,O^+_3, NF_3$
2	2	AB_3E_2	Angular V-Shaped	sp^3	$104^040^1, 102^0$	H2O, F2O
5	0	AB_5	Trigonal Bipyramid	sp^3d	$120^0, 90^0$	PCl5 , PF5
2	3	AB2 E3 (or) A_3E_3	Linear	sp^3d	180^0	XeF_2, I_3
6	0	AB_6	Octahedron	sp^3d^2	90^0	SF6

A = Central atom in the compound

B = atom linked to the central atom

E = lone pairs of electron.

◉ A covalent bond formed between two atoms the electron clouds concentrated along the internuclear axis having a cylindrical symmetry is known as a sigma bond.

◉ A covalent bond formed by a side wise overlap of p orbitals (or) p and d orbitals of atoms, in which the electron clouds are present above and below the internuclear axis is known as pi-bond.

◉ Strength of bonds $\sigma_{p-p} > \sigma_{s-s} > \pi_{p-p}$

- px and py - orbitals will not overlap because their orientations and symmetries are not same.

- The intermixing of atomic orbitals of almost equal energies of an atom and their redistribution into equal number of identical orbitals.

- In PCl5 the axal bonds of P-Cl are longer than equatorial bonds because the repulsions of the axial bond pairs stronger than equatorial bond pairs.

1.4.3 Molecular orbital theory

Region around the Nuclei of the bonded atoms in a molecule where the probability of finding electrons is maximum.

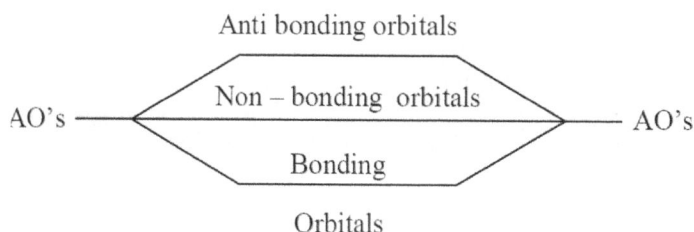

Order of energies

- **Bonding orbitals < Non - bonding orbitals < anti - bonding orbitals**

Sl. No.	AO of 1st atom	AO of 2nd atom	Designation of bonding Mo	Designation of Antibonding Mo
1	S	S or P_x	σ	$\sigma*$
2	P_x	S or P_x having end on over lap	σ	$\sigma*$
3	P_y	P_y side wise over lap	π	πP_y*
4	P_z	P_z side wise over lap	π	πP_z*

Sequence of energy levels of molecular orbitals

a) For lighter elements (B,C or N)

$$\sigma 1S < \sigma 1S* < \sigma 2S < \sigma*2S < \begin{Bmatrix} \pi 2P_y \\ \pi 2P_z \end{Bmatrix} < \sigma 2P_x < \begin{Bmatrix} \pi*2P_y \\ \pi*2P_z \end{Bmatrix} < \sigma*2P_x$$

b) For oxygen and heavier elements :

$$\sigma 1S < \sigma 1S* < \sigma 2S < \sigma*2S < \sigma 2P_x < \begin{Bmatrix} \pi 2P_y \\ \pi 2P_z \end{Bmatrix} < \begin{Bmatrix} \pi*2P_y \\ \pi*2P_z \end{Bmatrix} < \sigma*2P_x$$

$$\text{Bond order} = \frac{\text{Bonding electron} - \text{Antibonding electrons}}{2}$$

Hydrogen bond formed between the atoms of the same molecule is known as intramolecular hydrogen bond.

⊙ Hydrogen bond formed betwen two different polar molecules (the same (or) different substance) is known as inter molecular hydrogen bond.

⊙ As the temperature decreases thermal energy of gases decreases. The inter molecular repulsions decreases while attractive forces increases and therefore gases changes to liquid.

⊙ **Chemical Bond.** It is the force of attraction which holds or binds the constituent atoms together in a molecule.

⊙ **Octet Rule.** An atom tends to gain, lose or share electrons during molecule formation such that there are eight electrons surrounding it.

⊙ **Valence Electrons.** The electrons present in outermost shell are called valence electrons.

⊙ **Valence Bond Theory**

   ⅄ Covalent bond is formed by overlapping of half filled atomic orbitals.

   ⅄ Lowering of energy must take place during covalent bond formation. Attractive forces should be more than repulsive forces.

   ⅄ Strength of bond depends upon extent of overlapping. Greater the extend of overlapping stronger will be the bond.

⊙ **Ionic** Bond. It is the electrostatic force of attraction which holds the array of oppositely charged ions together. It is formed by complete transfer of electrons from one atom to another.

⊙ **Covalent Bond.** It is the force of attraction which arises due to the mutual sharing of electrons between two atoms. This definition is according to electronic theory proposed by G. N. Lewis in 1916.

⊙ **Lone Pair of Electrons.** The pair of electrons which do not take part in bond formation are called !one pair of electrons or non-bonding pair or unshared pairs.

⊙ **Double Bond.** When a bond is formed by sharing two electrons of each of the atom, it is called i double bond, *e.g.,*: O :: O : or : O = O:

⊙ **Triple Bond.** It is formed by sharing of three electrons of each atom, *e.g.,* N N or N N

⊙ **VSEPR Theory (Valence-Shell Electron- Pair RepulsionTheory).** According to this theory, the direction of the bonds around an atom in a molecule depends upon the total number of electron pairs (bonding as well as non-bonding) in the valence shell of an atom. Since the electron pairs repel each other, therefore that geometrical arrangement will be favoured in which repulsion is minimum.

◉ **Covalent Bond.** It is formed by overlapping of atomic orbitals having unpaired electrons according to theory based on modern wave mechanical model.

◉ **Bond Length.** It is defined as the average distance between the centres of the nuclei of the two bonded atoms in a molecule corresponding to minimum energy and maximum stability. It is measured in pm (picomctre).

◉ **Bond Energy.** The bond energy of a particular bond in a compound is defined as the average amount of energy produced when one mole of bonds is formed from isolated gaseous atoms or the amount of energy required when 1 mole of bonds is broken to get isolated gaseous atoms. It is measured in kilo Joules.

◉ **Bond Angle.** It is the average angle between two (molecular orbitals) bonds meeting at the same nucleus in a poly-atomic molecule. It is measured in degrees and minutes.

◉ **Hybridization.** It is the phenomenon of intermixing of atomic orbitals of slightly different energies to give rise to new hybridized orbitals having equivalent energy and identical shapes.

◉ Hybridized orbitals are more stable than atomic orbitals.

◉ **sp3 Hybridization.** If one s and three p-orbitals are intermixed or hybridized, this is known as sp3-hybridization. Four equivalent orbitals will he formed which are called sp3 hybridized orbitals which are directed along the four corners of a tetrahedron with bond angle 109° 28', *e.g.,* in CH4, carbon undergoes sp3 hybridization and it has tetrahedral shape.

◉ **sp3d Hybridisation.** When one *s*, three *p* and one d-orbitals intermix together and give rise to five sp3d hybridised orbitals. The shape of molecule is *trigonal bipyramidal.* Bond angle is 90°, 120°, e.g., PF3, PCl3 involves . sp3dhybridisation.

◉ **sp^3 d2 Hybridisation.** When one *s,* three *p* and two d-orbitals intermix to give rise to six sp^{3}d^2-hybridised orbitals. The shape of molecule having.sp3 *d2* hybridisation will be octahedral with bond angle 90°, *e.g.,*SF6[CrF6]2– involve sp3d2 hybridisation and possess octahedral shape.

◉ **sp2d or dsp2 Hybridisation.** When one *s,* two p and one d-orbitals intermix to give rise to four sp2d or *dsp2* hybridised orbitals having square planar shape with bond angle 90°, *e.g.,* [Ni (CN)4]2– undergo dsp2 hybridisation and are square planar in shape.

◉ **sp^2 Hybridization.** If one s and two p-orbitals are intermixed, it is called sp2-hybridization. Three equivalent orbitals are called sp2 hybridized orbitals which will be directly along the corners of equilateral triangle having bond angle 120°, *e.g.,* in BF3, 'B' atom undergoes sp2 hybridization and its shape is a planar.

◉ **sp Hybridization.** If one s and one *p* orbital intermix, it is called sp-hybridization. The two *sp* hybridized orbitals are tar-apart. In BeF2. Be atom undergoes sp-hybridization and it has linear shape with bond angle 180°.

◉ **Sigma Bond.** The bond which is formed by the head on overlapping of half-filled atomic orbitals along the inter-nuclear axis, is called sigma bond.

◉ **Pi Bond.** The bond which is formed by the sideways or lateral overlapping of half-filled atomic orbitals in a direction perpendicular to the inter-nuclear axis, is called pi bond.

◉ **The Coordinate Covalent Bond.** It is a special type of covalcnt bond in which shared pair of electrons is supplied only by one of the atoms forming the bond. The atom which supplies the electrons is called donor while the other atom which only uses the shared pair of electrons is known as acceptor. It is shown by arrow-head (®)

◉ **Polar Covalent Bond,** if covalent bond is formed between two dissimilar atoms having difference in electronegativity. then the shared pair of electrons is more towards more electronegative atom which acquired d – charge and other atom acquires d +| charge. This type of bond is called polar covalent bond. The compounds having polar covalent bond form ions in aqueous solution. It is also called ionic character of the covalent bond.

◉ **Dipole Moment.** Dipole moment of a molecule is the product of its net positive or negative charge and distance between the atoms. Mathematically, m = $e \times d$, where 2 = net + ve or -ve charge, d is distance between the atoms. Its unit is Debye, where 1 Debye = 1018 e.g. u. cm. Its S.I. unit is Cm (Coulomb meter).

◉ **Ionic Solids.** The solids in which there is three dimensional array of cations and anions are called ionic solids, *e.g.,* LiCl, LiF, NaCl, NaF, etc.

◉ **Covalent Solids.** In covalent solids, atoms are linked to one another by covalent bonds forming a giant network, e.g., SiC (silicon carbide), diamond are examples of covalent solids.

◉ **Metallic Bond.** It is formed by the simultaneous force of attraction between the Kernels (+vely charged atoms) and mobile electrons which hold the metal atoms together. All metals and alloys have metallic bond.

◉ **Resonance.** In case of some of the molecules, all of their properties can't be explained by a single structure. The molecule can be represented by two or more structures. These structures are called canonical forms. The molecule is said to be resonance hybrid of all these structures. These structures have no physical reality in the sense that they cannot be prepared in the laboratory. They exist in imagination. This phenomenon of existing in more than one structure is called resonance, e.g.,.

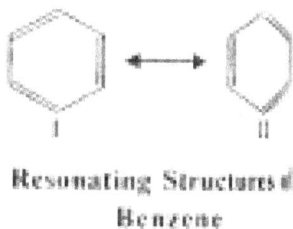

Resonating Structures of
Benzene

⊙ **Resonance Energy.** It is defined as difference in energy of actual structure of compound and most stable resonating structure, *e.g.,* benzene has resonance energy 150,3 kJ mol–1 or 36 kcal mol^{-1}.

⊙ **Predicting the Hybrid State of Atom in Different Species:**

⋏ Write the valence electrons of central atom(V).

⋏ Add number of surrounding atoms except oxygen (S A).

⋏ If there is positive charge on species, subtract the charge, if there is -ve charge, add the charge.

⋏ Divide the sum by 2 to get value of X.c

$$X = \frac{1}{2}(V + SA \pm E)$$

Value of X	2	3	4	5	6	7
Hybrid Siate of Central Atom	sp	sp^2	sp^3	sp^{3}d	sp^{3}d^2	sp^{3}d^2

Note. The above formula is rot applicable to predict hybrid state of metal in complexes and species

⊙ **Molecular Orbital Theory.** Molecular orbital gives electron probability distribution around a group of nuclei in a molecule. They are filled in the same way as atomic orbitals. Molecular

⊙ orbitals are formed by linear combination of atomic orbitals.

⊙ l **Bonding molecular orbital.** A molecular orbital that *is* formed by addition overlap (*i.e.,*

⊙ when the lobes of atomic orbitals overlap with the same sign) of two atomic orbitals is known as *bonding molecular orbital.* It is represented as AB A B y = y + y Its energy is lower than the atomic orbitals from which it is formed. It favours bonding.

- **Anti-bonding molecular orbital.** A molecular orbital that is obtained by the subtraction overlap (*i.e.,* when the lobes of atomic orbitals overlap with the opposite sign) of two atomic orbitals is known as *anti-bonding molecular orbital.* It is represented as *AB A B y = y - y

It is energy is higher than the atomic orbitals from which it is formed. It does not favour bonding.

- **Bond order. It** is defined as half of the difference between number of electrons in bonding and anti-bonding orbitals, *i.e.,* Bond order = 1/2 (Nb - Na) where Nb are number of electrons in bonding orbitals and Na are number of electrons in anti-bonding orbitals. Bond order helps in estimating stability of atom.

- **Relationship between electronic configuration and molecular behaviour:**
  - ⊀ If Nb is greater than Na, the molecule is stable.
  - ⊀ The molecule is unstable if Na, is greater than Nb.

The molecule is also unstable if Na is equal to Nb because anti-bonding effect is stronger than bonding effect.

Sigma (s) molecular orbitals. A molecular orbital which is formed from the overlap of two. atomic orbitals or head to head overlap of one *s* and p-atomic orbitals or head to head overlap

of two p-atomic orbitals, is known as *sigma molecular orbital.*

- ⊀ **Pi (p) molecular orbitals.** A molecular orbital which is formed by lateral overlap of two parallel *p*-orbitals is known as *pi (p) molecular orbital.*

- ⊀ **Paramagnetism.** The presence of one or more unpaired electrons is the cause of paramagnetism. A paramagnetic substance is attracted by an applied magnetic field.

- ⊀ **Diamagnetism.** The absence of unpaired electrons is the cause of diamagnetism. A diamagnetic substance will be repelled by an applied magnetic field.

- ⊀ **Ferromagnetism.** It is due to large number of unpaired electrons. The substance remains magnetised even after applied magnetic field is removed.

- **Conditions for the Combination of Atomic Orbitals.** The linear combination of atomic orbitals takes place only if the following conditions are satisfied: (/) The combining atomic orbitals must have same or nearly same energy. It means 1s orbiial combines with 1s orbital but not with 2s orbital because energy of 2s orbital is appreciably higher than 1s orbital. 2s orbital can overlap with 2s orbital to form s 2s and s 2s as shown in. Figure.

⊙ **Energy level Diagrams for Molecular Orbitals.** We have observed that 1s atomic orbitals on two atoms form two molecular orbitals designated as σ1s and a"s *1s. In the same manner, the *2s* and *2p* atomic orbitals (eight atomic orbitals) give rise to the following eights molecular orbitals.

Antibonding	MO's	s *2s z	s *2p x	p *2p y	p * 2p
Bonding	MO's	s 2s z	s 2p x	p2p y	p2p

⊙ The energy levels of these molecular orbitals have been determined experimentally from the spectroscopic data for homonuclear diatomic molecules of second row (period) elements of periodic table.

The increasing order of energies of various molecular orbitals for O2 and F2 is given below.

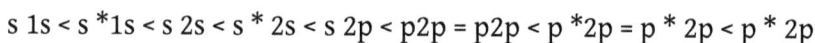

$$s\ 1s < s\ *1s < s\ 2s < s\ *\ 2s < s\ 2p < p2p = p2p < p\ *2p = p\ *\ 2p < p\ *\ 2p$$

⊙ However, this sequence of energy levels of molecular orbitals is not correct for remaining mofeculi Li2, Be2, B2, C2, N2. For instance, it has been observed experimentally that for molecules such as B2, C2, N2 etc., the increasing order of energies of various molecular orbitals is

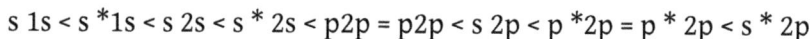

$$s\ 1s < s\ *1s < s\ 2s < s\ *\ 2s < p2p = p2p < s\ 2p < p\ *2p = p\ *\ 2p < s\ *\ 2p$$

The important characteristic feature of this order is that the energy of z s 2p **molecular orbital is night than that of** x p2p **and** y p2p **molecular orbitals** in these molecules

The Hydrogen Bond

It is intermolecular force of attraction between hydrogen atom of one molecule and a highly electronegative element (such as N, O or F) within the same molecule or another molecule of same or different compounds.

The H atom attached to nitrogen, oxygen or fluorine is able to interpose itself between two such electronegative atoms bonding them together. This is called hydrogen bond.

The essential requirements for an H-bond are :

(i) a hydrogen atom attached to highly electronegative atom

(ii) a lone pair of electrons on electronegative atom

The hydrogen bond is written as

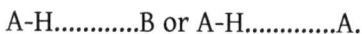

A-H............B or A-H.............A.

Table 1. Some Homunuclear molecules and their ions, Heteronuclear molecules and ions

S. No.	Molecule/ion	No. of electrons	Electronic Configuration	Bond Order $\frac{1}{2}(N_b - N_a)$	Magnetic Property
1.	H_2	2	$(\sigma 1s)^2$	$\frac{1}{2}(2-0)=1$	Diamagnetic
2.	H_2^+	1	$(\sigma 1s)^1$	$\frac{1}{2}(1-0)=1/2$	Paramagnetic
3.	H_2^-	3	$(\sigma 1s)^2(\sigma^*1s)^1$	$\frac{1}{2}(2-1)=1/2$	Paramagnetic
4.	He_2	4	$(\sigma 1s)^2(\sigma^*1s)^2$	$\frac{1}{2}(2-2)=0$	Does not exist
5.	He_2^+	3	$(\sigma 1s)^2(\sigma^*1s)^1$	$\frac{1}{2}(2-1)=1/2$	Paramagnetic
6.	Li_2	6	$(\sigma 1s)^2(\sigma^*1s)^2(\sigma 2s)^2$	$\frac{1}{2}(4-2)=1$	Diamagnetic
7.	Be_2	8	$(\sigma 1s)^2(\sigma^*1s)^2(\sigma 2s)^2(\sigma^*2s)^2$	$\frac{1}{2}(4-4)=0$	Does not exist
8.	B_2	10	$(\sigma 1s)^2(\sigma^*1s)^2(\sigma 2s)^2(\sigma^*2s)^2(\pi 2p_x^1 \pi 2p_y^1)$	$\frac{1}{2}(6-4)=1$	Paramagnetic
9.	C_2	12	$(\sigma 1s)^2(\sigma^*1s)^2(\sigma 2s)^2(\sigma^*2s)^2(\pi 2p_x^2 \pi 2p_y^2)$	$\frac{1}{2}(8-4)=2$	Diamagnetic
10.	C_2^+	11	$(\sigma 1s)^2(\sigma^*1s)^2(\sigma 2s)^2(\sigma^*2s)^2(\pi 2p_x^2 \pi 2p_y^1)$	$\frac{1}{2}(7-4)=3/2$	Paramagnetic
11.	C_2^-	13	$(\sigma 1s)^2(\sigma^*1s)^2(\sigma 2s)^2(\sigma^*2s)^2(\pi 2p_x^2 \pi 2p_y^1)(\sigma 2p_z^1)$	$\frac{1}{2}(9-4)=5/2$	Paramagnetic
12.	N_2	14	$(\sigma 1s)^2(\sigma^*1s)^2(\sigma 2s)^2(\sigma^*2s)^2(\pi 2p_x^2 \pi 2p_y^2)(\sigma 2p_z^2)$	$\frac{1}{2}(10-4)=3$	Diamagnetic
13.	N_2^+	13	$(\sigma 1s)^2(\sigma^*1s)^2(\sigma 2s)^2(\sigma^*2s)^2(\pi 2p_x^2 \pi 2p_y^2)(\sigma 2p_z^1)$	$\frac{1}{2}(9-4)=5/2$	Paramagnetic
14.	O_2	16	$(\sigma 1s)^2(\sigma^*1s)^2(\sigma 2s)^2(\sigma^*2s)^2(\sigma 2p_z^2)(\pi 2p_x^2 \pi 2p_y^2)(\pi^*2p_x^1 \pi^*2p_y^1)$	$\frac{1}{2}(10-6)=2$	Paramagnetic
15.	O_2^+	15	$(\sigma 1s)^2(\sigma^*1s)^2(\sigma 2s)^2(\sigma^*2s)^2(\sigma 2p_z^2)(\pi 2p_x^2 \pi 2p_y^2)(\pi^*2p_x^1)$	$\frac{1}{2}(10-5)=5/2$	Paramagnetic
16.	O_2^-	17	$(\sigma 1s)^2(\sigma^*1s)^2(\sigma 2s)^2(\sigma^*2s)^2(\sigma 2p_z^2)(\pi 2p_x^2 \pi 2p_y^2)(\pi^*2p_x^2\pi^*2p_y^1)$	$\frac{1}{2}(10-7)=3/2$	Paramagnetic
17.	O_2^{2-}	18	$(\sigma 1s)^2(\sigma^*1s)^2(\sigma 2s)^2(\sigma^*2s)^2(\sigma 2p_z^2)(\pi 2p_x^2 \pi 2p_y^2)(\pi^*2p_x^2 \pi^*2p_y^2)$	$\frac{1}{2}(10-8)=1$	Diamagnetic
18.	F_2	18	$(\sigma 1s)^2(\sigma^*1s)^2(\sigma 2s)^2(\sigma^*2s)^2(\sigma 2p_z^2)(\pi 2p_x^2 \pi 2p_y^2)(\pi^*2p_x^2 \pi^*2p_y^2)$	$\frac{1}{2}(10-8)=1$	Diamagnetic
19.	F_2^+	17	$(\sigma 1s)^2(\sigma^*1s)^2(\sigma 2s)^2(\sigma^*2s)^2(\sigma 2p_z^2)(\pi 2p_x^2 \pi 2p_y^2)(\pi^*2p_x^2 \pi^*2p_y^1)$	$\frac{1}{2}(10-7)=3/2$	Paramagnetic
20.	F_2^-	19	$(\sigma 1s)^2(\sigma^*1s)^2(\sigma 2s)^2(\sigma^*2s)^2(\sigma 2p_z^2)(\pi 2p_x^2 \pi 2p_y^2)(\pi^*2p_x^2 \pi^*2p_y^2)(\sigma^*2p_z^1)$	$\frac{1}{2}(10-9)=1/2$	Paramagnetic
21.	Ne_2	20	$(\sigma 1s)^2(\sigma^*1s)^2(\sigma 2s)^2(\sigma^*2s)^2(\sigma 2p_z^2)(\pi 2p_x^2 \pi 2p_y^2)(\pi^*2p_x^2 \pi^*2p_y^2)(\sigma^*2p_z^2)$	$\frac{1}{2}(10-10)=0$	Does not exist
22.	Ne_2^+	19	$(\sigma 1s)^2(\sigma^*1s)^2(\sigma 2s)^2(\sigma^*2s)^2(\sigma 2p_z^2)(\pi 2p_x^2\pi 2p_y^2)(\pi^*2p_x^2 \pi^*2p_y^2)(\sigma^*2p_z^1)$	$\frac{1}{2}(10-9)=1/2$	Paramagnetic
23.	NO	15	$(\sigma 1s)^2(\sigma^*1s)^2(\sigma 2s)^2(\sigma^*2s)^2(\sigma 2p_z^2)(\pi 2p_x^2 \pi 2p_y^2)(\pi^*2p_x^1)$	$\frac{1}{2}(10-5)=5/2$	Paramagnetic
24.	$(NO)^+$	14	$(\sigma 1s)^2(\sigma^*1s)^2(\sigma 2s)^2(\sigma^*2s)^2(\sigma 2p_z^2)(\pi 2p_x^2 \pi 2p_y^2)$	$\frac{1}{2}(10-4)=3$	Diamagnetic
25.	CO	14	$(\sigma 1s)^2(\sigma^*1s)^2(\sigma 2s)^2(\sigma^*2s)^2(\sigma 2p_z^2)(\pi 2p_x^2 \pi 2p_y^2)$	$\frac{1}{2}(10-4)=3$	Diamagnetic

02 | IONIC BONDING

2.1 INTRODUCTION: AN OVERVIEW

Ionic bonding is a fundamental concept in chemistry that underpins the formation of countless compounds, from common table salt to intricate mineral structures. At its core, ionic bonding represents the dynamic interplay between positively and negatively charged ions, resulting in the creation of stable chemical compounds. This captivating phenomenon occurs when atoms, driven by the desire to attain a stable electron configuration, either gain or lose electrons to achieve a noble gas configuration. Through this process, they transform into ions, which then engage in electrostatic interactions, akin to a cosmic dance of attraction and repulsion, to form robust ionic compounds.

The allure of ionic bonding lies not only in its simplicity but also in its profound implications for understanding the properties and behaviors of substances in the natural world. It serves as the cornerstone of various disciplines, from materials science to biochemistry, offering insights into the structures of minerals, the conductivity of solutions, and the intricacies of biological processes. Moreover, the principles of ionic bonding extend far beyond the confines of laboratory experiments, manifesting in everyday phenomena that shape our lives and surroundings.

In this exploration of ionic bonding, we embark on a journey to unravel its mysteries, delving into the mechanisms that govern the formation of ionic compounds and the factors that influence their properties. From the timeless principles articulated by early pioneers to the cutting-edge discoveries fueling contemporary research, we traverse the vast landscape of chemical bonding, seeking to grasp the essence of one of nature's most captivating phenomena. Through experimentation, observation, and theoretical abstraction, we endeavor to illuminate the intricate tapestry of interactions that define the realm of ionic bonding and its profound significance in the grand narrative of chemistry.

2.2 FORMATION OF IONIC BONDS

Ionic bonds are characterized by the complete transfer of electrons from one atom to another, resulting in the formation of two charged particles known as ions, which are held together with the help of electrostatic forces. An Ionic bond is formed when one of the atoms can donate electrons to achieve the inert gas electron configuration and the other atom needs electrons to achieve the inert gas electron configuration. That is the chemical bond formed by the transfer of electrons from one atom to another. Ionic bond is also known as electrovalent bond and compounds composed of Ionic bonds are called ionic compounds. Therefore, it will not be wrong to say that ions form ionic compounds. When a metal reacts with a non-metal, then they form ionic bond and the compound is called the ionic compound. As a result of reaction between metal and non-metal, they are bonded with electrostatic force of attraction with each other; such bonds are called chemical bonds.

Ionic bonds form through the transfer of electrons from one atom to another. This transfer occurs between atoms with significantly different electronegativities, resulting in the creation of ions: one atom loses electrons to become a positively charged cation, while the other gains electrons to become a negatively charged anion. The attraction between these oppositely charged ions then leads to the formation of an ionic bond. This type of bonding typically occurs between metals and non-metals, where the metal tends to lose electrons (forming a positive ion) and the non-metal tends to gain electrons (forming a negative ion). The resulting compound is held together by strong electrostatic forces between the ions, creating a stable structure known as an ionic lattice.

For an example:

Formation of Sodium Chloride Sodium is a metal whereas chlorine is a non-metal. Sodium metal reacts with chlorine to form an ionic compound, sodium chloride. Now we will see how sodium chloride is formed and what changes takes place in the electronic arrangements of sodium and chlorine atoms in the formation of this compound.

The atomic number of sodium is 11, so its electronic configuration is $1s^2$,$2s^2$,$2p^6$,$3s^1$. Sodium has only one electron in its outermost shell. So, sodium atom will donate one electron to chlorine atom and forms a sodium ion i.e. Na^+ .

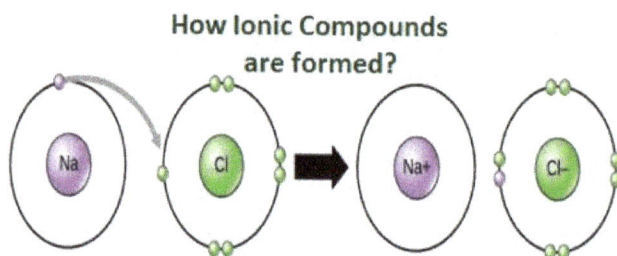

How Ionic Compounds
are formed?

On the other side chlorine atomic number is 17, its electronic configuration is $1s^2$,$2s^2$,$2p^6$,$3s^2$, $3p^5$. Therefore, chlorine atom has 7 electrons in the outermost shell and needs one more electron to achieve stable electronic configuration or inert gas configuration. So, a chlorine atom takes one electron from sodium atom and forms a negatively charged chloride ion i.e. Cl⁻.

2.3 PROPERTIES OF IONIC COMPOUNDS

Physical Properties of Ionic Compounds

Melting Points: Because of the many simultaneous attractions between cations and anions that occur, ionic crystal lattices are very strong. The process of melting an ionic compound requires the addition of large amounts of energy in order to break all of the ionic bonds in the crystal. For example, sodium chloride has a melting temperature of about 800°C .

Shattering: Ionic compounds are generally hard, but brittle. Why? It takes a large amount of mechanical force, such as striking a crystal with a hammer, to force one layer of ions to shift relative to its neighbor. However, when that happens, it brings ions of the same charge next to each other. The repulsive forces between like-charged ions cause the crystal to shatter. When an ionic crystal breaks, it tends to do so along smooth planes because of the regular arrangement of the ions.

Conductivity: The ionic bonded molecules in their aqueous solutions or in the molten state are good conductors of electricity. This is due to the presence of ions which acts as charge carriers.

Ionic Bonding and Lattice Energy

Ions are atoms or molecules which are electrically charged. Cations are positively charged and anions are negatively charged. Ions form when atoms gain or lose valence electrons. Since electrons are negatively charged, an atom that loses one or more electrons will become positively charged; an atom that gains one or more electrons becomes negatively charged. Ionic bonding is the attraction between positively- and negatively-charged ions. These oppositely charged ions attract each other to form ionic networks, or lattices. Electrostatics explains why this happens: opposite charges attract and like charges repel. When many ions attract each other, they form large, ordered, crystal lattices in which each ion is surrounded by ions of the opposite charge. Generally, when metals react with non-metals, electrons are transferred from the metals to the non-metals. The metals form positively-charged ions and the nonmetals form negatively-charged ions.

The properties of ionic compounds follow from the orderly crystal lattice arrangement of tightly bonded charged particles that make them up. Ionic compounds tend to have high melting and boiling points, because the attraction between ions in the lattice is very strong. Moving ions out of the lattice disrupts the structure, so ionic compounds tend to be brittle rather than malleable. Ionic compounds do not conduct electricity in the solid state because ions are not free to move around the lattice; however, when ionic compounds are dissolved, they may dissociate into individual ions which move freely through the solution and therefore conduct electricity well.

Generating Ionic Bonds

Ionic bonds form when metals and non-metals chemically react. By definition, a metal is relatively stable if it loses electrons to form a complete valence shell and becomes positively charged. Likewise, a nonmetal becomes stable by gaining electrons to complete its valence shell and become negatively charged. When metals and non-metals react, the metals lose electrons by transferring them to the non-metals, which gain them. Consequently, ions are formed, which instantly attract each other—ionic bonding. In the overall ionic compound, positive and negative charges must be balanced, because electrons cannot be created or destroyed, only transferred. Thus, the total number of electrons lost by the cationic species must equal the total number of electrons gained by the anionic species.

Ionic compounds are held together by electrostatic forces, which are described in classical physics by Coulomb's Law. According to this law, the energy of the electrostatic attraction (E) between two charged particles is proportional to the magnitude of the charges Q1 and Q2 and inversely proportional to the internuclear distance between the particles (r):

$$E \propto Q_1 Q_2 / r$$

The energy of attraction (E) is a type of potential energy, since it is based on the position of the charged particles relative to each other. If the two particles have opposite charges (as in ionic compounds), the value of (EE) will be negative, meaning that energy is released by bringing the particles together—that is, the particles naturally attract each other. According to Coulomb's Law, the larger the magnitude of the charges on each particle, the stronger the attraction will be. So, for example, Mg^{2+} and O^{2-} will have a stronger attraction than Na^+ and Cl^-, because of the larger charges. Also, the closer together the charges are, the stronger the attraction. Therefore, smaller ions also form stronger ionic bonds.

In an ionic lattice, many more than two charged particles interact simultaneously, releasing an amount of energy known as the lattice energy. The lattice energy is not exactly the same as that predicted by Coulomb's Law, but the same general principles of electrostatic attraction apply. In an ionic compound, the value of the lattice energy corresponds to the strength of the ionic bonding.

Electron Configuration of Ions

If ionic bonding becomes stronger for compounds with more highly charged ions, why does sodium only lose one electron to form Na+ rather than, say, Na²⁺? The number of electrons transferred between ions depends not only on the energy released in lattice formation but also on the energy required to strip away electrons from one atom and add them to another. In other words, the lattice energy released by ionic compound formation must be balanced against the required ionization energy and electron affinity of forming the ions. Since the Na+ ion has a noble gas electron configuration, stripping away the next electron from this stable arrangement would require more energy than what is released during lattice formation (Sodium I2 = 4,560 kJ/mol). Thus, sodium is present in ionic compounds as Na$^+$ and not Na^{2+}. Likewise, adding an electron to fill a valence shell (and achieve noble gas electron configuration) is exothermic or only slightly endothermic. To add an additional electron into a new subshell requires tremendous energy - more than the lattice energy. Thus, we find Clin ionic compounds, but not Cl^{2-}. As a general rule, main group elements only form ions with the nearest noble gas electron configuration - otherwise, the lattice energy would not be enough to compensate for the ionization energy / electron affinity.

Typical values of lattice energy can compensate for values as large as I$_3$ for valence electrons (i.e. can strip away up to 3 valence electrons from cations). Because most transition metals would require the removal of more than 3 electrons to attain a noble gas core, they are not found in ionic compounds with a noble gas core. A transition metal always loses electrons first from the higher 's' subshell, before losing from the underlying 'd' subshell. (The remaining electrons in the unfilled d subshell are the reason for the bright colors observed in many transition metal compounds!) For example, iron ions will not form a noble gas core:

- Fe: $[Ar]4s^2 3d^6$

- Fe^{2+}: $[Ar] 3d^6$

- Fe^{3+}: $[Ar] 3d^5$

Some metal ions can form a pseudo noble gas core (and be colorless), for example:

- Ag: $[Kr]5s^1 4d^{10}$ Ag$^+$ $[Kr]4d^{10}$ Compound: AgCl

- Cd: $[Kr]5s^2 4d^{10}$ Cd^{2+} $[Kr]4d^{10}$ Compound: CdS

Note: The silver and cadmium atoms lost the 5s electrons in achieving the ionic state. Remember that atoms always lost electrons from the subshell with the highest n quantum number first (i.e. 5s before 4d). When a positive ion is formed from an atom, electrons are always lost first from the subshell with the largest principle quantum number.

Polyatomic Ions

Not all ionic compounds are formed from only two elements. Many polyatomic ions exist, in which two or more atoms are bound together by covalent bonds. They form a stable grouping which carries a charge (positive or negative). The group of atoms as a whole act as a charged species in forming an ionic compound with an oppositely charged ion. Polyatomic ions may be either positive or negative, for example:

- NH_4^+ (ammonium) = cation
- SO_4^{2-} (sulfate) = anion

The principles of ionic bonding with polyatomic ions are the same as those with monatomic ions. Oppositely charged ions come together to form a crystalline lattice, releasing a lattice energy. Based on the shapes and charges of the polyatomic ions, these compounds may form crystalline lattices with interesting and complex structures.

Energetics of Ionic Bond Formation

Ionic bonds are formed when positively and negatively charged ions are attracted by electrostatic forces. Consider a single pair of ions, one cation and one anion. How strong will the force of their attraction be? We can rewrite Coulomb's Law quantitatively for any two charged particles:

$$E=kQ_1Q_2/r$$

where each ion's charge is represented by the symbol Q and the internuclear distance between the particles is represented by (rr). The proportionality constant k is equal to 2.31×10^{-28} J·m. This value of k includes the charge of a single electron (1.6022×10^{-19} C) for each ion. The equation can also be written using the charge of each ion, expressed in coulombs (C), incorporated in the constant. In this case, the proportionality constant, k, equals 8.999×109 J·m/C^2 . In the example given, Q_1 = +1(1.6022×10^{-19} C) and Q_2 = -1(1.6022 × 10-19 C). If Q1 and Q2 have opposite signs (as in NaCl, for example, where Q_1 is +1 for Na$^+$ and Q_2 is -1 for Cl$^-$), then E is negative, which means that energy is released when oppositely charged ions are brought together from an infinite distance to form an isolated ion pair.

Energy is always released when a bond is formed and correspondingly, it always requires energy to break a bond.

As shown by the green curve in the lower half of below Figure , the maximum energy would be released when the ions are infinitely close to each other, at r = 0. Because ions occupy space and have a structure with the positive nucleus being surrounded by electrons, however, they cannot be infinitely close together. At very short distances, repulsive electron–electron interactions between electrons on adjacent ions become stronger than the attractive interactions between ions with opposite charges, as shown by the red curve in the upper half of Figure . The total energy of the system is a balance between the attractive and repulsive interactions. The purple curve in Figure shows that the total energy of the system reaches a minimum at r0, the point where the electrostatic repulsions and attractions are exactly balanced. This distance is the same as the experimentally measured bond distance.

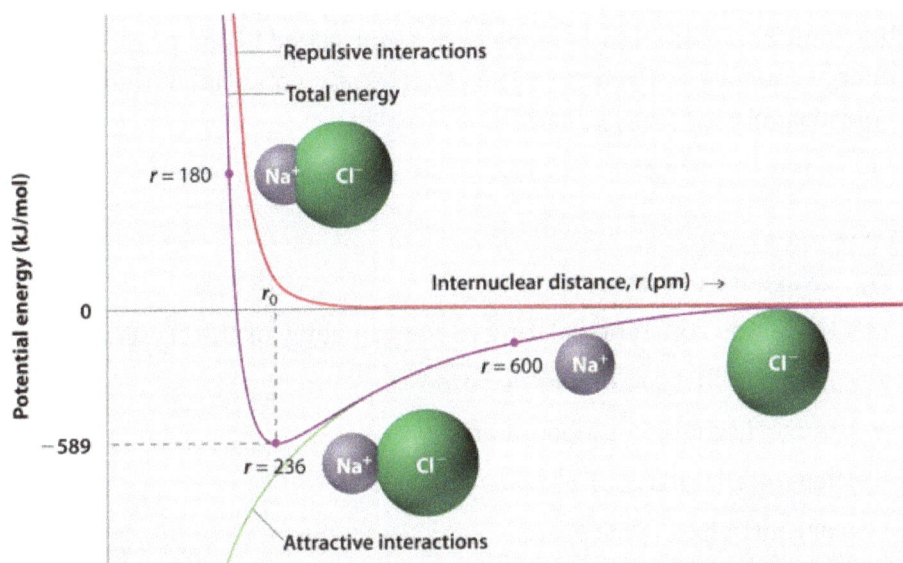

Figure: Plot of Potential Energy versus Internuclear Distance for the Interaction between a Gaseous Na⁺ Ion and a Gaseous Cl⁻ Ion. The energy of the system reaches a minimum at a particular distance (r_0) when the attractive and repulsive interactions are balanced.

Consider the energy released when a gaseous Na+ ion and a gaseous Cl⁻ ion are brought together from $r = \infty$ to $r = r_0$. Given that the observed gas-phase internuclear distance is 236 pm, the energy change associated with the formation of an ion pair from an Na⁺ (g) ion and a Cl⁻ (g) ion is as follows:

$E = kQ1Q2/r_0$

$= (2.31 \times 10^{-28} \text{J·m})[(+1)(-1)/236\text{pm} \times 10^{-12}\text{m/pm}]$

$= -9.79 \times 10^{-19}$ J/ion pair

The negative value indicates that energy is released. Our convention is that if a chemical process provides energy to the outside world, the energy change is negative. If it requires energy, the energy change is positive. To calculate the energy change in the formation of a mole of NaCl pairs, we need to multiply the energy per ion pair by Avogadro's number:

E = (−9.79×10⁻¹⁹ J/ion pair) (6.022×10²³ ion pair/mol) = −589kJ/mol

This is the energy released when 1 mol of gaseous ion pairs is formed, not when 1 mol of positive and negative ions condenses to form a crystalline lattice. Because of long-range interactions in the lattice structure, this energy does not correspond directly to the lattice energy of the crystalline solid.

Born−Landé equation

The **Born−Landé equation** is a mean of calculating the lattice energy of a crystalline ionic compound. In 1918 Max Born and Alfred Landé proposed that the lattice energy could be derived from the electrostatic potential of the ionic lattice and a repulsive potential energy term.

$$\Delta U(0K) = -N_A M |Z^+||Z^-| e^2 (1-1/n)/4\pi\epsilon_0 r_0$$

where:

- N_A = Avogadro constant;
- M = Madelung constant, relating to the geometry of the crystal;
- z+ = numeric charge number of cation
- z− = numeric charge number of anion
- e = elementary charge, 1.6022×10^{-19} C
- ε0 = permittivity of free space

$4\pi\epsilon0 = 1.112\times10^{-10}$ C²/(J·m)

Derivation

The ionic lattice is modeled as an assembly of hard elastic spheres which are compressed together by the mutual attraction of the electrostatic charges on the ions. They achieve the observed equilibrium distance apart due to a balancing short-range repulsion.

Electrostatic potential

The electrostatic potential energy, $E_{pair,}$ between a pair of ions of equal and opposite charge is:

$$\Delta U = -Z^2 e^2 /4\pi\epsilon_0 r$$

where

z = magnitude of charge on one ion

e = elementary charge, 1.6022×10^{-19} C

ϵ_0 = permittivity of free space

$4\pi\epsilon_0 = 1.112\times10^{-10}$ C²/(J·m)

r = distance separating the ion centers

For a simple lattice consisting ions with equal and opposite charge in a 1:1 ratio, interactions between one ion and all other lattice ions need to be summed to calculate EM, sometimes called the Madelung or lattice energy:

$$\Delta U = - N_A M Z^2 e^2 / 4\pi \epsilon_0 r$$

where

M = Madelung constant, which is related to the geometry of the crystal

r = closest distance between two ions of opposite charge

Repulsive term

Born and Lande suggested that a repulsive interaction between the lattice ions would be proportional to $1/r^n$ so that the repulsive energy term, ER, would be expressed:

$$\Delta U = N_A B / r^n$$

where

B = constant scaling the strength of the repulsive interaction

r = closest distance between two ions of opposite charge

n = Born exponent, a number between 5 and 12 expressing the steepness of the repulsive barrier

The total intensive potential energy of an ion in the lattice can therefore be expressed as the sum of the Madelung and repulsive potentials:

$$\Delta U = -M N_A Z^2 e^2 / 4\pi \epsilon_0 r + N_A B / r^n$$

Minimizing this energy with respect to r yields the equilibrium separation r_0 in terms of the unknown constant B

$$B = - N_A M Z^2 e^2 r_0^{(n-1)} / 4\pi \epsilon_0 n$$

Evaluating the minimum intensive potential energy and substituting the expression for B in terms of r0 yields the Born–Landé equation:

$$E(r_0) = -M N_A Z^2 e^2 (1-1/n) / 4\pi \epsilon_0 r_0$$

Solvation energy

The change in Gibbs energy when an ion or molecule is transferred from a vacuum (or the gas phase) to a solvent. The main contributions to the solvation energy come from:

- ⊙ The cavitation energy of formation of the hole which preserves the dissolved species in the solvent;

- ⊙ The orientation energy of partial orientation of the dipoles;

- The isotropic interaction energy of electrostatic and dispersion origin; and
- The anisotropic energy of specific interactions, e.g. hydrogen bonds, donor-acceptor interactions etc.

Lattice Energy

Lattice Energy is a type of potential energy that may be defined in two ways. In one definition, the lattice energy is the energy required to break apart an ionic solid and convert its component atoms into gaseous ions. This definition causes the value for the lattice energy to always be positive, since this will always be an endothermic reaction. The other definition says that lattice energy is the reverse process, meaning it is the energy released when gaseous ions bind to form an ionic solid. As implied in the definition, this process will always be exothermic, and thus the value for lattice energy will be negative. Its values are usually expressed with the units kJ/mol.

Lattice Energy is used to explain the stability of ionic solids. Some might expect such an ordered structure to be less stable because the entropy of the system would be low. However, the crystalline structure allows each ion to interact with multiple oppositely charge ions, which causes a highly favorable change in the enthalpy of the system. A lot of energy is released as the oppositely charged ions interact. It is this that causes ionic solids to have such high melting and boiling points. Some require such high temperatures that they decompose before they can reach a melting and/or boiling point.

2.4 BORN-HABER CYCLE

There are several important concepts to understand before the Born-Haber Cycle can be applied to determine the lattice energy of an ionic solid; ionization energy, electron affinity, dissociation energy, sublimation energy, heat of formation, and Hess's Law.

- **Ionization Energy** is the energy required to remove an electron from a neutral atom or an ion. This process always requires an input of energy, and thus will always have a positive value. In general, ionization energy increases across the periodic table from left to right, and decreases from top to bottom. There are some excepts, usually due to the stability of half-filled and completely filled orbitals.

- **Electron Affinity** is the energy released when an electron is added to a neutral atom or an ion. Usually, energy released would have a negative value, but due to the definition of electron affinity, it is written as a positive value in most tables. Therefore, when used in calculating the lattice energy, we must remember to subtract the electron affinity, not add it. In general, electron affinity increases from left to right across the periodic table and decreases from top to bottom.

- **Dissociation energy** is the energy required to break apart a compound. The dissociation of a compound is always an endothermic process, meaning it will always require an input of energy. Therefore, the change in energy is always positive. The magnitude of the dissociation energy depends on the electronegativity of the atoms involved.

- **Sublimation energy** is the energy required to cause a change of phase from solid to gas, bypassing the liquid phase. This is an input of energy, and thus has a positive value. It may also be referred to as the energy of atomization.

- **The heat of formation** is the change in energy when forming a compound from its elements. This may be positive or negative, depending on the atoms involved and how they interact.

- **Hess's Law** states that the overall change in energy of a process can be determined by breaking the process down into steps, then adding the changes in energy of each step. The Born-Haber Cycle is essentially Hess's Law applied to an ionic solid.

2.3.1 Using the Born-Haber Cycle

The values used in the Born-Haber Cycle are all predetermined changes in enthalpy for the processes described in the section above. Hess' Law allows us to add or subtract these values, which allows us to determine the lattice energy.

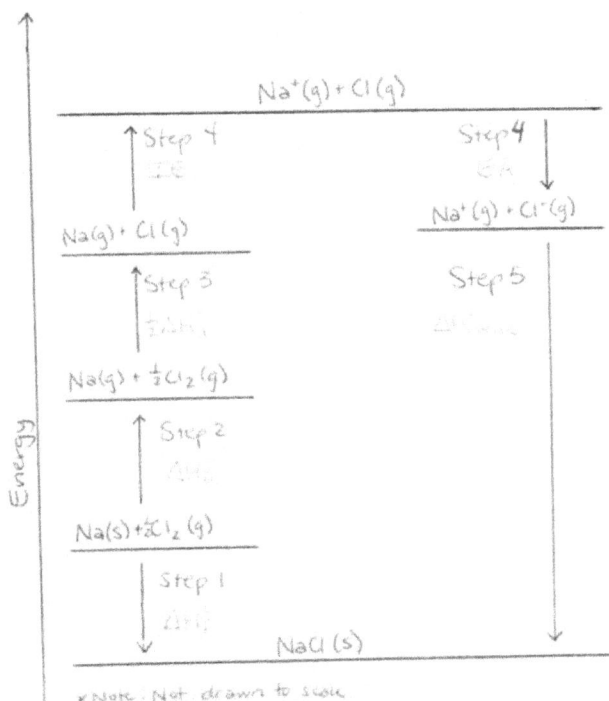

Step 1: Determine the energy of the metal and nonmetal in their elemental forms. (Elements in their natural state have an energy level of zero.) Subtract from this the heat of formation of the ionic solid that would be formed from combining these elements in the appropriate ration. This is the energy of the ionic solid, and will be used at the end of the process to determine the lattice energy.

Step 2: The Born-Haber Cycle requires that the elements involved in the reaction are in their gaseous forms. Add the changes in enthalpy to turn one of the elements into its gaseous state, and then do the same for the other element.

Step 3: Metals exist in nature as single atoms and thus no dissociation energy needs to be added for this element. However, many nonmetals will exist as polyatomic species. For example, Cl exists as Cl2 in its elemental state. The energy required to change Cl2 into 2Cl atoms must be added to the value obtained in Step 2.

Step 4: Both the metal and nonmetal now need to be changed into their ionic forms, as they would exist in the ionic solid. To do this, the ionization energy of the metal will be added to the value from Step 3. Next, the electron affinity of the nonmetal will be subtracted from the previous value. It is subtracted because it is a release of energy associated with the addition of an electron.

*This is a common error due to confusion caused by the definition of electron affinity, so be careful when doing this calculation.

Step 5: Now the metal and nonmetal will be combined to form the ionic solid. This will cause a release of energy, which is called the lattice energy. The value for the lattice energy is the difference between the value from Step 1 and the value from Step 4.

The diagram below is another representation of the Born-Haber Cycle.

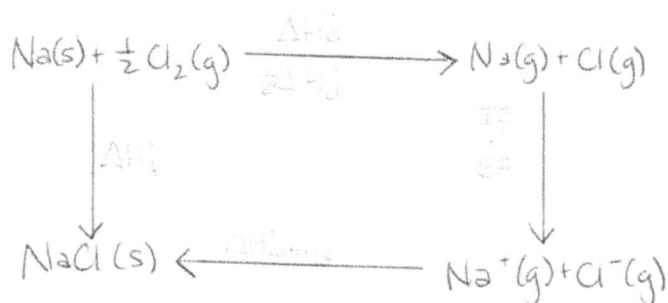

Equation

The Born-Haber Cycle can be reduced to a single equation:

Heat of formation= Heat of atomization+ Dissociation energy+ (sum of Ionization energies)+ (sum of Electron affinities)+ Lattice energy

*Note: In this general equation, the electron affinity is added. However, when plugging in a value, determine whether energy is released (exothermic reaction) or absorbed (endothermic reaction) for each electron affinity. If energy is released, put a negative sign in front of the value; if energy is absorbed, the value should be positive.

Rearrangement to solve for lattice energy gives the equation:

Lattice energy= Heat of formation- Heat of atomization- Dissociation energy- (sum of Ionization energies)- (sum of Electron Affinities)

Lattice enthalpy is a measure of the strength of the forces between the ions in an ionic solid. The greater the lattice enthalpy, the stronger the forces.

Defining Lattice Enthalpy

There are two different ways of defining lattice enthalpy which directly contradict each other, and you will find both in common use. In fact, there is a simple way of sorting this out, but many sources do not use it. Lattice enthalpy is a measure of the strength of the forces between the ions in an ionic solid. The greater the lattice enthalpy, the stronger the forces. Those forces are only completely broken when the ions are present as gaseous ions, scattered so far apart that there is negligible attraction between them. You can show this on a simple enthalpy diagram.

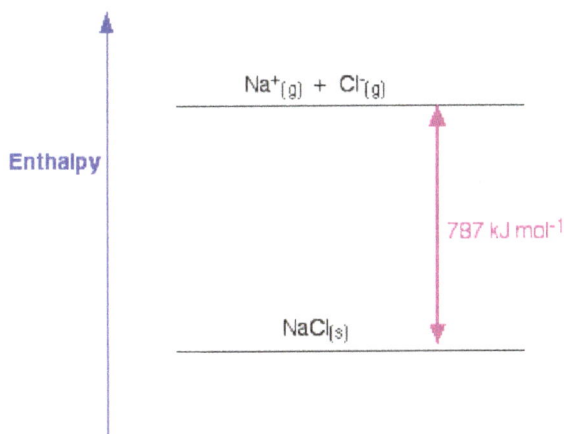

For sodium chloride, the solid is more stable than the gaseous ions by 787 kJ mol-1, and that is a measure of the strength of the attractions between the ions in the solid. Remember that energy (in this case heat energy) is released when bonds are made, and is required to break bonds.

So lattice enthalpy could be described in either of two ways.

⊙ It could be described as the enthalpy change when 1 mole of sodium chloride (or whatever) was formed from its scattered gaseous ions. In other words, you are looking at a downward arrow on the diagram.

- Or, it could be described as the enthalpy change when 1 mole of sodium chloride (or whatever) is broken up to form its scattered gaseous ions. In other words, you are looking at an upward arrow on the diagram.

Both refer to the same enthalpy diagram, but one looks at it from the point of view of making the lattice, and the other from the point of view of breaking it up. Unfortunately, both of these are often described as "lattice enthalpy".

Definitions

- The lattice dissociation enthalpy is the enthalpy change needed to convert 1 mole of solid crystal into its scattered gaseous ions. Lattice dissociation enthalpies are always positive.

- The lattice formation enthalpy is the enthalpy change when 1 mole of solid crystal is formed from its separated gaseous ions. Lattice formation enthalpies are always negative.

This is an absurdly confusing situation which is easily resolved by never using the term "lattice enthalpy" without qualifying it.

- You should talk about "lattice dissociation enthalpy" if you want to talk about the amount of energy needed to split up a lattice into its scattered gaseous ions. For NaCl, the lattice dissociation enthalpy is +787 kJ mol^{-1}.

- You should talk about "lattice formation enthalpy" if you want to talk about the amount of energy released when a lattice is formed from its scattered gaseous ions. For NaCl, the lattice formation enthalpy is -787 kJ mol^{-1}.

That immediately removes any possibility of confusion.

Factors affecting Lattice Enthalpy

The two main factors affecting lattice enthalpy are

- The charges on the ions and
- The ionic radii (which affects the distance between the ions).

The charges on the ions

Sodium chloride and magnesium oxide have exactly the same arrangements of ions in the crystal lattice, but the lattice enthalpies are very different.

We can see that the lattice enthalpy of magnesium oxide is much greater than that of sodium chloride. That's because in magnesium oxide, 2+ ions are attracting 2- ions; in sodium chloride, the attraction is only between 1+ and 1- ions.

Lattice enthalpies of NaCl and MgO

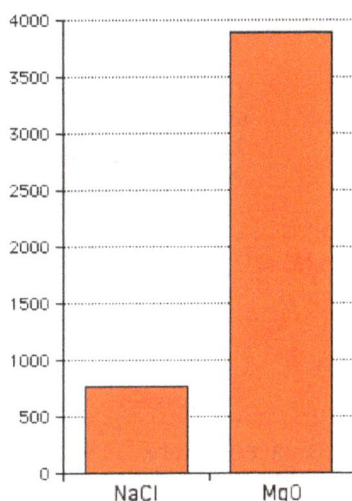

The bar chart shows NaCl at approximately 750 and MgO at approximately 3900, with the vertical axis ranging from 0 to 4000 in increments of 500.

The Radius of the Ions

The lattice enthalpy of magnesium oxide is also increased relative to sodium chloride because magnesium ions are smaller than sodium ions, and oxide ions are smaller than chloride ions. That means that the ions are closer together in the lattice, and that increases the strength of the attractions.

This effect of ion size on lattice enthalpy is clearly observed as you go down a Group in the Periodic Table. For example, as you go down Group 7 of the Periodic Table from fluorine to iodine, you would expect the lattice enthalpies of their sodium salts to fall as the negative ions get bigger - and that is the case:

Lattice enthalpies of sodium halides

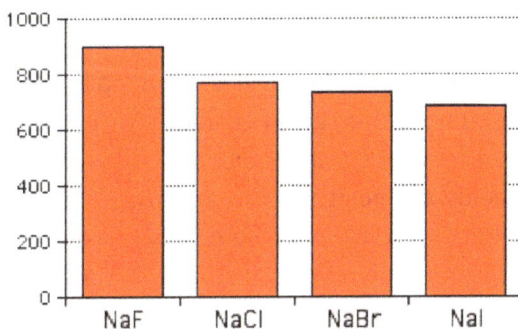

The bar chart shows NaF at approximately 900, NaCl at approximately 780, NaBr at approximately 740, and NaI at approximately 690, with the vertical axis ranging from 0 to 1000 in increments of 200.

Attractions are governed by the distances between the centers of the oppositely charged ions, and that distance is obviously greater as the negative ion gets bigger. And you can see exactly the same effect if as you go down Group 1. The next bar chart shows the lattice enthalpies of the Group 1 chlorides.

Lattice enthalpies of Group 1 chlorides

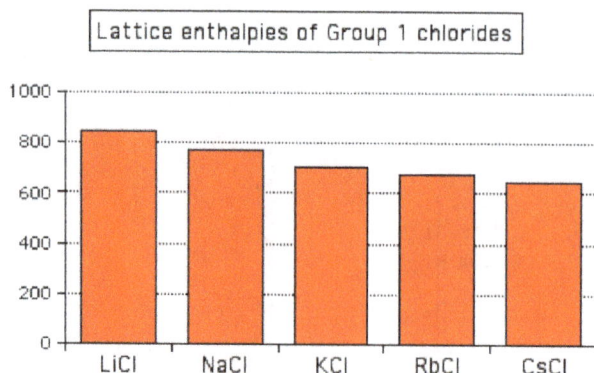

Calculating Lattice Enthalpy

It is impossible to measure the enthalpy change starting from a solid crystal and converting it into its scattered gaseous ions. It is even more difficult to imagine how you could do the reverse - start with scattered gaseous ions and measure the enthalpy change when these convert to a solid crystal. Instead, lattice enthalpies always have to be calculated, and there are two entirely different ways in which this can be done.

1.   You can can use a Hess's Law cycle (in this case called a Born-Haber cycle) involving enthalpy changes which can be measured. Lattice enthalpies calculated in this way are described as experimental values.

2.   Or you can do physics-style calculations working out how much energy would be released, for example, when ions considered as point charges come together to make a lattice. These are described as theoretical values. In fact, in this case, what you are actually calculating are properly described as lattice energies.

Born-Haber Cycles

Standard Atomization Enthalpies

Before we start talking about Born-Haber cycles, we need to define the atomization enthalpy, ΔHoa . The standard atomization enthalpy is the enthalpy change when 1 mole of gaseous atoms is formed from the element in its standard state. Enthalpy change of atomization is always positive. You are always going to have to supply energy to break an element into its separate gaseous atoms. All of the following equations represent changes involving atomization enthalpy:

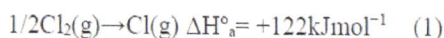

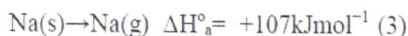

$$1/2Cl_2(g) \rightarrow Cl(g) \ \Delta H°_a = +122 kJmol^{-1} \quad (1)$$

$$12Br_2(l) \rightarrow Br(g) \ \Delta H°_a = +122 kJmol^{-1} \quad (2)$$

$$Na(s) \rightarrow Na(g) \ \Delta H°_a = +107 kJmol^{-1} \quad (3)$$

Notice particularly that the "mol^{-1}" is per mole of atoms formed - NOT per mole of element that you start with. You will quite commonly have to write fractions into the left-hand side of the equation. Getting this wrong is a common mistake.

2.3.2 BORN-HABER CYCLE FOR NaCl

Consider a Born-Haber cycle for sodium chloride, and then talk it through carefully afterwards. You will see that I have arbitrarily decided to draw this for lattice formation enthalpy. If you wanted to draw it for lattice dissociation enthalpy, the red arrow would be reversed - pointing upwards.

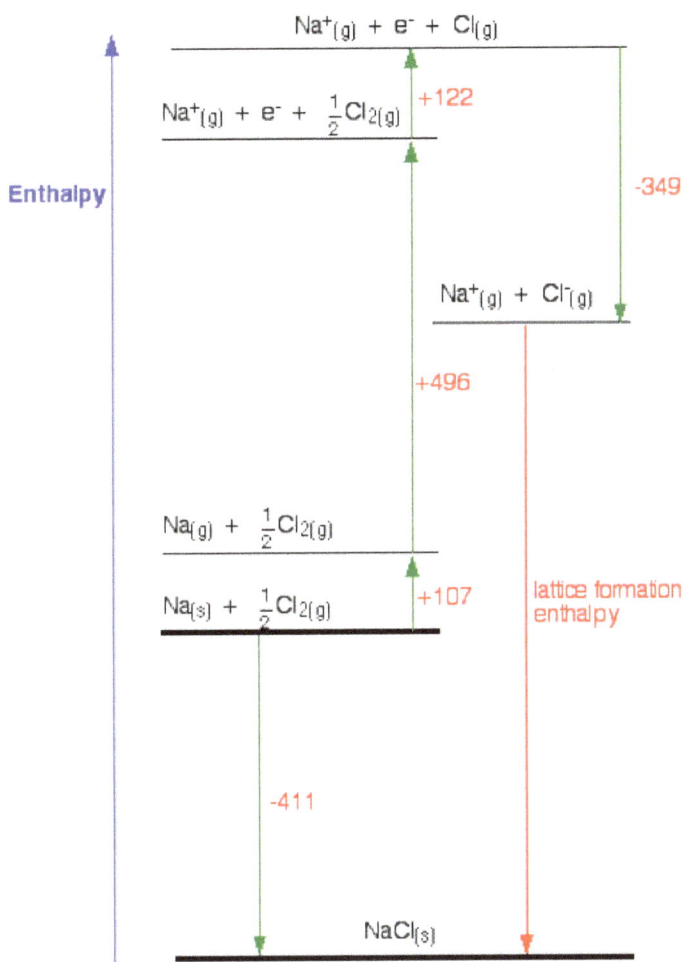

Focus to start with on the higher of the two thicker horizontal lines. We are starting here with the elements sodium and chlorine in their standard states. Notice that we only need half a mole of chlorine gas in order to end up with 1 mole of NaCl. The arrow pointing down from this to the lower thick line represents the enthalpy change of formation of sodium chloride.

The Born-Haber cycle now imagines this formation of sodium chloride as happening in a whole set of small changes, most of which we know the enthalpy changes for - except, of course, for the lattice enthalpy that we want to calculate.

- ⊙ The +107 is the atomization enthalpy of sodium. We have to produce gaseous atoms so that we can use the next stage in the cycle.

- ⊙ The +496 is the first ionization energy of sodium. Remember that first ionization energies go from gaseous atoms to gaseous singly charged positive ions.

- ⊙ The +122 is the atomization enthalpy of chlorine. Again, we have to produce gaseous atoms so that we can use the next stage in the cycle.

- ⊙ The -349 is the first electron affinity of chlorine. Remember that first electron affinities go from gaseous atoms to gaseous singly charged negative ions.

- ⊙ And finally, we have the positive and negative gaseous ions that we can convert into the solid sodium chloride using the lattice formation enthalpy.

Now we can use Hess' Law and find two different routes around the diagram which we can equate. As drawn, the two routes are obvious. The diagram is set up to provide two different routes between the thick lines. So, from the cycle we get the calculations directly underneath it . .

$-411 = +107 + 496 + 122 - 349 + LE$

$LE = -411 - 107 - 496 - 122 + 349$

$LE = -787 \text{ kJ mol}^{-1}$

How would this be different if you had drawn a lattice dissociation enthalpy in your diagram? Your diagram would now look like this:

The only difference in the diagram is the direction the lattice enthalpy arrow is pointing. It does, of course, mean that you have to find two new routes. You cannot use the original one, because that would go against the flow of the lattice enthalpy arrow. This time both routes would start from the elements in their standard states, and finish at the gaseous ions.

$-411 + LE = +107 + 496 + 122 - 349$

$LE = +107 + 496 + 122 - 349 + 411$

$LE = +787 \text{ kJ mol}^{-1}$

Once again, the cycle sorts out the sign of the lattice enthalpy.

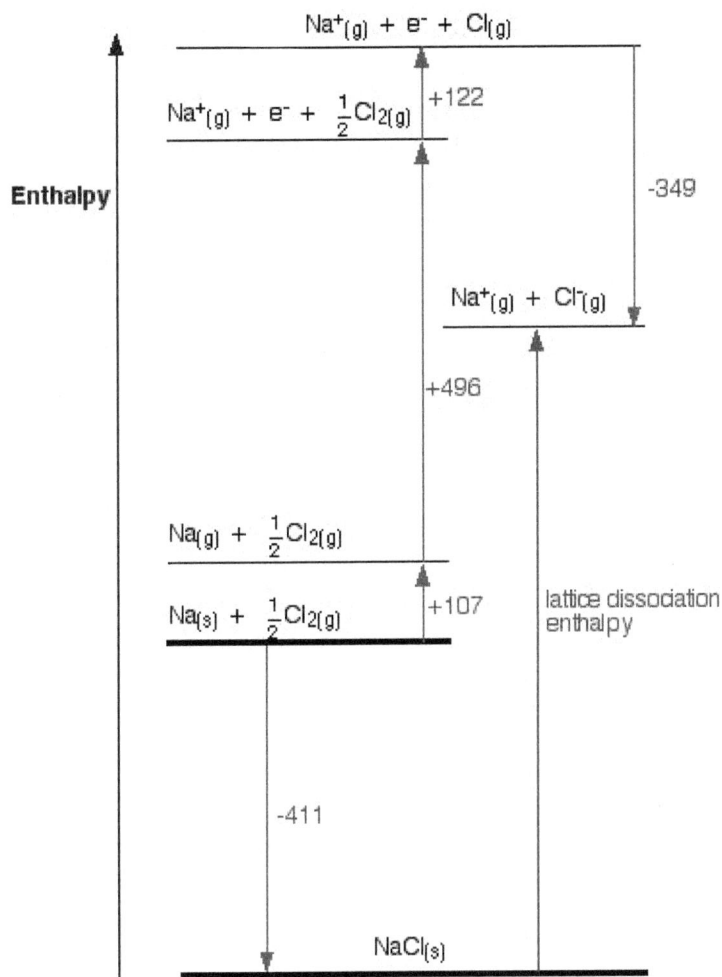

$$Na^+_{(g)} + e^- + Cl_{(g)}$$

$$Na^+_{(g)} + e^- + \tfrac{1}{2}Cl_{2(g)} \quad +122$$

Enthalpy

$$-349$$

$$Na^+_{(g)} + Cl^-_{(g)}$$

$$+496$$

$$Na_{(g)} + \tfrac{1}{2}Cl_{2(g)}$$

$$Na_{(s)} + \tfrac{1}{2}Cl_{2(g)} \quad +107 \quad \text{lattice dissociation enthalpy}$$

$$-411$$

$$NaCl_{(s)}$$

Lattice Energies and Solubility

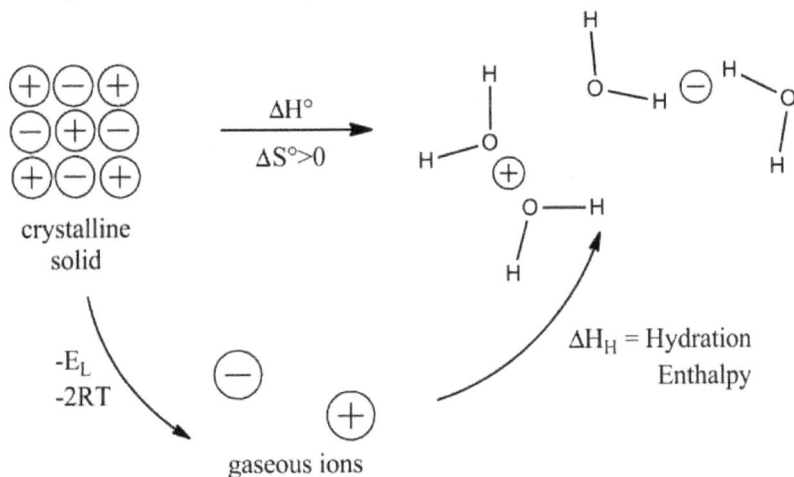

$\Delta H°$

$\Delta S°>0$

crystalline solid

$-E_L$
$-2RT$

gaseous ions

ΔH_H = Hydration Enthalpy

Lattice energies can also help predict compound solubilities. Let's consider a Born-Haber cycle for dissolving a salt in water. We can imagine this as the sum of two processes: (1) the vaporization of the salt to produce gaseous ions, characterized by the lattice enthalpy, and (2) the hydration of those ions to produce the solution. The enthalpy change for the overall process is the sum of those two steps. We know that the entropy change for dissolution of a solid is positive, so the solubility depends on the enthalpy change for the overall process.

Here we need to consider the trends in both the lattice energy EL and the hydration energy EH. The lattice energy depends on the sum of the anion and cation radii $(r+ + r-)$, whereas the hydration energy has separate anion and cation terms. Generally the solvation of small ions (typically cations) dominates the hydration energy because of the $1/r2$ dependence.

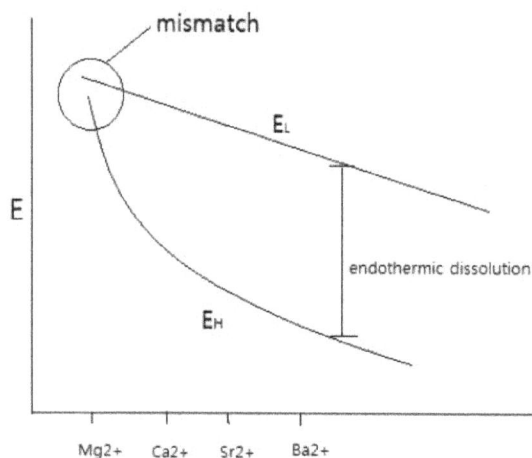

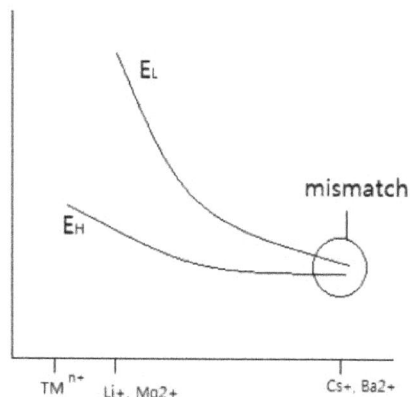

Figure A

Figure B

Left: EL diagram for sulfate salts. The large SO_4^{2-} ion is size-mismatched to small cations such as Mg^{2+}, which have large hydration soluble salts. With larger cations such as Ba^{2+}, which have lower EH, the lattice energy exceeds the solvation enthalpy and the the case of small anions such as F^- and OH^-, the lattice energy dominates with small cations such as transition metal ions (TMcation size mismatch occurs with larger cations, such as Cs^+ and Ba^{2+}, which make soluble fluoride salts.

For small anions, EL is more sensitive to $r+$, whereas EH does not depend on $r+$ as strongly. For fluorides and hydroxides, LiF is slightly soluble whereas CsF is very soluble, and $Mg(OH)2$ is insoluble whereas $Ba(OH)2$ is very soluble.

Putting both trends together, we see that **low solubility** is most often encountered when the **anion and cation match well in their sizes**, especially when one or both are **multiply charged**.

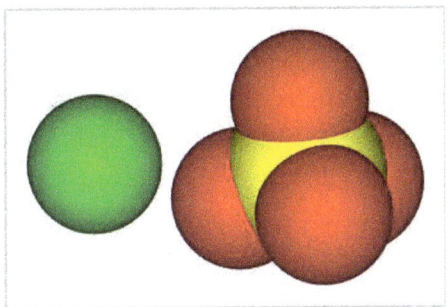

Space-filling models showing the van der Waals surfaces of Ba2+ and SO42- of BaSO4 in water.

Combining all our conclusions about solubility, we note the following trends:

1. Increasing **size mismatch** between the anion and cation leads to greater solubility, so CsF and LiI are the most soluble alkali halides.

2. Increasing **covalency** leads to lower solubility in the salts (due to larger EL. For example, AgF, AgCl, AgBr, and AgI exhibit progressively lower solubility because of increasing covalency.

$$AgF>AgCl>AgBr>AgIAgF>AgCl>AgBr>AgI$$

3. Increasing the **charge on the anion** lowers the solubility because the increase in EL is large relative to the increase in EH.

4. Small, polyvalent cations (having large EH) make **soluble salts with large, univalent anions** such as I^-, NO_3^-, ClO_4^-, PF_6^-, and acetate.

Examples: Salts of transition metal and lanthanide ions

⊙ Ln^{3+}: Nitrate salts are soluble, but oxides and hydroxides are insoluble.

⊙ Fe^{3+}: Perchlorate is soluble, but sulfate is insoluble.

5. Multiple charged anions such as O_2^-, S_2^-, PO_4^{3-}, and SO_4^{2-} make insoluble salts with most M^{2+}, M^{3+}, and M^{4+} metals.

2.5 IONIC RADIUS TRENDS AND PAULING'S RULES

Ionic radius trends and Pauling's rules are foundational concepts in the study of chemical bonding, particularly in understanding the behavior of ions in ionic compounds. Ionic radius refers to the size of an ion, typically defined as the distance from the nucleus to the outermost electron shell. This radius plays a crucial role in determining the overall structure, stability, and properties of ionic compounds.

The concept of ionic radius trends revolves around the systematic variation in the size of ions within a given group or period of the periodic table. Generally, as one moves down a group in the periodic table, the ionic radius of the ions increases due to the addition of electron shells. Conversely, along a period, the ionic radius tends to decrease as the effective nuclear charge increases, pulling the electrons closer to the nucleus.

Pauling's rules, formulated by Linus Pauling in the early 20th century, provide guidelines for predicting the crystal structures of ionic compounds based on the sizes and charges of ions. These rules are instrumental in understanding the arrangements of ions in solid-state structures and are invaluable in materials science and crystallography.

Pauling's First Rule states that in a stable ionic compound, the coordination number (the number of nearest neighbor ions) of the cation and anion should be as high as possible to maximize the strength of the electrostatic interactions. This rule helps in predicting the coordination geometry of ions in crystal lattices.

Pauling's Second Rule addresses the principle of electrostatic neutrality, asserting that the sum of the strengths of the electrostatic forces acting on an anion from all surrounding cations should be equal to the sum of the strengths of the electrostatic forces acting on a cation from all surrounding anions. This rule aids in predicting the stoichiometry of ionic compounds and understanding the arrangement of ions in crystal structures.

Finally, Pauling's Third Rule relates to the sharing of edges and faces in crystal lattices to minimize repulsive interactions between ions of the same charge, thereby stabilizing the crystal structure.

Together, ionic radius trends and Pauling's rules provide a robust framework for understanding the behavior of ions in ionic compounds, elucidating the factors that govern their structures, stabilities, and properties. They serve as indispensable tools in the exploration of chemical bonding and the design of novel materials with tailored functionalities and characteristics.

2.4.1 Ionic Radii and Radius Ratios

Atoms in crystals are held together by electrostatic forces, van der Waals interactions, and covalent bonding. It follows that arrangements of atoms that can maximize the strength of these attractive interactions should be most favorable and lead to the most commonly observed crystal structures.

Radius ratio rules

Early crystallographers had trouble solving the structures of inorganic solids using X-ray diffraction because some of the mathematical tools for analyzing the data had not yet been developed. Once a trial structure was proposed, it was relatively easy to calculate the diffraction pattern, but it was difficult to go the other way (from the diffraction pattern to the structure) if nothing was known *a priori* about the arrangement of atoms in the unit cell. It was (and still is!) important to develop some guidelines for guessing the coordination numbers and bonding geometries of atoms in crystals. The first such rules were proposed by Linus Pauling, who considered how one might pack together oppositely charged spheres of different radii. Pauling proposed from geometric considerations that the quality of the "fit" depended on the **radius ratio** of the anion and the cation.

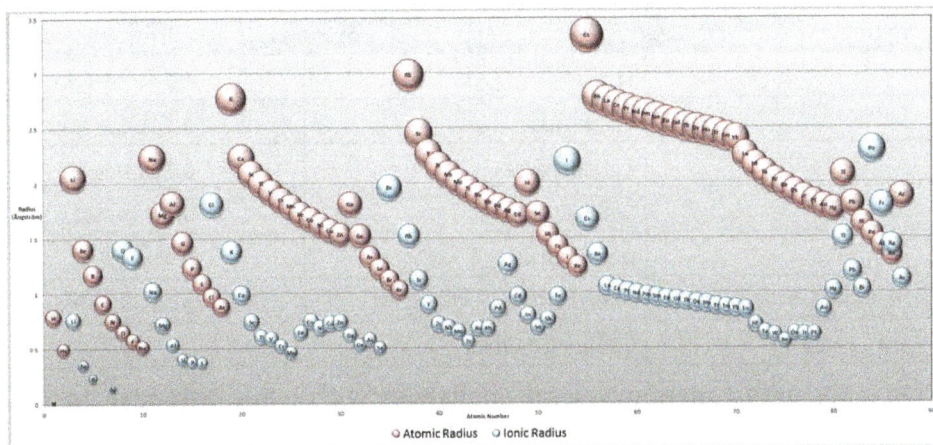

Atomic Radius   Ionic Radius

2.4.2 Atomic and Ionic Radii.

Note that cations are always smaller than the neutral atom (pink) of the same element, whereas anions are larger.

Going from left to right across any row of the periodic table, neutral atoms and cations contract in size because of increasing nuclear charge.

The basic idea of radius ratio rules is illustrated at the right. We consider that the anion is the packing atom in the crystal and the smaller cation fills interstitial sites ("holes"). Cations will find arrangements in which they can contact the largest number of anions. If the cation can touch all of its nearest neighbor anions, as shown at the right for a small cation in contact with larger anions, then the fit is good. If the cation is too small for a given site, that coordination number will be unstable and it will prefer a lower coordination structure. The table below gives the ranges of cation/anion radius ratios that give the best fit for a given coordination geometry.

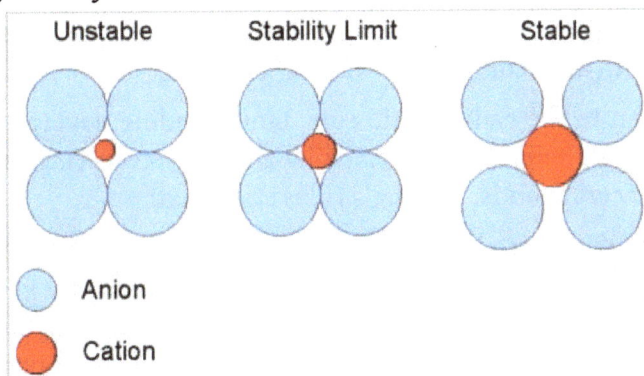

Unstable   Stability Limit   Stable

Anion

Cation

Critical Radius Ratio. This diagram is for coordination number six: 4 anions in the plane are shown, 1 is above the plane and 1 is below. The stability limit is at $r_c/r_A = 0.414$

Coordination number	Geometry	$\rho = r_{cation}/r_{anion}$
2	Linear	0 - 0.155
3	Triangular	0.155-0.225
4	Tetrahedral	0.225-0.414
4	Square Plannar	0.414-0.732
6	Octahedral	0.414-0.732
8	Cubic	0.732-1.000
12	Cubooctahedral	1.0

There are unfortunately several challenges with using this idea to predict crystal structures:

◉ We don't know the radii of individual ions

◉ Atoms in crystals are not really ions - there is a varying degree of covalency depending electronegativity differences

◉ Bond distances (and therefore ionic radii) depend on bond strength and coordination number (remember Pauling's rule $D(n) = D(1) - 0.6 \log n$)

◉ Ionic radii depend on oxidation state (higher charge => smaller cation size, larger anion size)

What about the alkali halides NaCl, KBr, LiI, CsF, etc.? All of them have the NaCl structure except for CsCl, CsBr, and CsI, which have the CsCl (8-8) structure. In this case the radius ratio model fails rather badly. The Li+ salts LiBr and LiI are predicted to have tetrahedral structures, and KF is predicted to have an 8-8 structure like CsCl. We can try adjusting the radii (e.g., making the cations larger and anions smaller), but the best we can do with the alkali halides is predict about half of their structures correctly. Since the alkali halides are clearly ionic compounds, this failure suggests that there is something very wrong with the radius ratio model, and its success with MO_2 compounds was coincidental.

◉ In addition to the radius ratio rule, Linus Pauling developed other useful rules that are helpful in rationalizing and also predicting the structures of inorganic compounds. Pauling's rules state that:

◉ Stable structures are **locally electroneutral**. For example, in the structure of the double perovskite Sr_2FeMoO_6, MO_6 (M = Fe^{2+}, Mo^{6+}) octahedra share all their vertices, and Sr^{2+} ions fill the cubooctahedral cavities that are flanked by eight MO_6 octahedra. Each O_2- ion is coordinated to one $Fe2+$

and one Mo6+ ion in order to achieve local electroneutrality, and thus the FeO6 and MoO6 octahedra alternate in the structure.

⊙ **Cation-cation repulsion** should be minimized. Anion polyhedra can share vertices (as in the perovskite structure) without any energetic penalty. Shared polyhedral edges, and especially shared faces, cause cation-cation repulsion and should be avoided. For example, in rutile, the most stable polymorph of $TiO2$, the TiO6 octahedra share vertices and two opposite edges, forming ribbons in the structure. In anatase $TiO2$, each octahedron shares four edges so the anatase polymorph is less thermodynamically stable.

⊙ Highly charged cations in anion polyhedra tend not to share edges or even vertices, especially when the coordination number is low. For example, in orthosilicates such as olivine (M_2SiO_4), there are isolated SiO_4^{4-} tetrahedra.

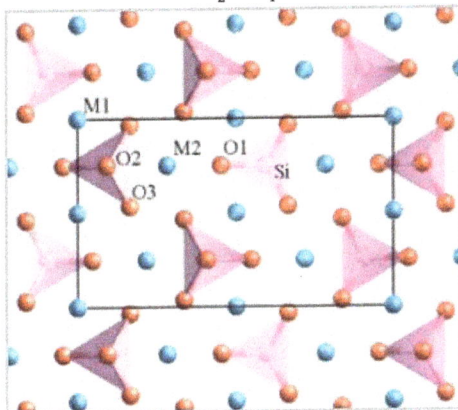

Structure of olivine. M (Mg or Fe) = blue spheres, Si = pink tetrahedra, O = red spheres.

As we will soon see, all of Pauling's rules are justified on the basis of lattice energy considerations. In ionic compounds, the arrangement of atoms that maximizes anion-cation interactions while minimizing cation-cation and anion-anion contacts is energetically the best.

Polarizability

Polarizability allows us to better understand the interactions between nonpolar atoms and molecules and other electrically charged species, such as ions or polar molecules with dipole moments.

Introduction

Neutral nonpolar species have spherically symmetric arrangements of electrons in their electron clouds. When in the presence of an electric field, their electron clouds can be distorted. The ease of this distortion is defined as the **polarizability** of the atom or molecule. The created distortion of the electron cloud causes the originally nonpolar molecule or atom to acquire a dipole moment. This induced dipole moment is related to the polarizability of the molecule or atom and the strength of the electric field by the following equation:

$$\mu\text{ind} = \alpha E \quad (1)$$

where E denotes the strength of the electric field and α is the polarizability of the atom or molecule with units of $C\ m^2 V^{-1}$

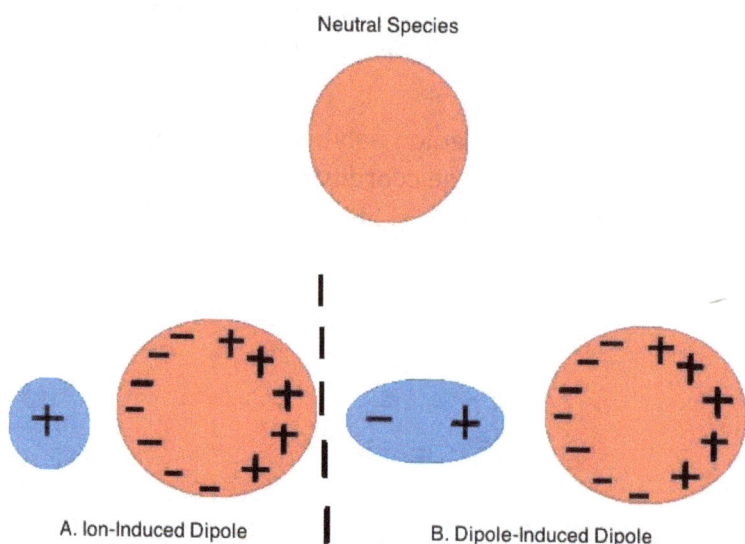

Neutral Species

A. Ion-Induced Dipole   B. Dipole-Induced Dipole

In general, polarizability correlates with the interaction between electrons and the nucleus. The amount of electrons in a molecule affects how tight the nuclear charge can control the overall charge distribution. Atoms with fewer electrons will have smaller, denser electron clouds, as there is a strong interaction between the few electrons in the atoms' orbitals and the positively charged nucleus. There is also less shielding in atoms with fewer electrons contributing to the stronger interaction of the outer electrons and the nucleus. With the electrons held tightly in place in these smaller atoms, these atoms are typically not easily polarized by external electric fields. In contrast, large atoms with many electrons, such as negative ions with excess electrons, are easily polarized. These atoms typically have very diffuse electron clouds and large atomic radii that limit the interaction of their external electrons and the nucleus.

2.4.3 Factors that Influence Polarizability

The relationship between polarizability and the factors of electron density, atomic radii, and molecular orientation is as follows:

- ◉ The greater the number of electrons, the less control the nuclear charge has on charge distribution, and thus the increased polarizability of the atom.

- The greater the distance of electrons from nuclear charge, the less control the nuclear charge has on the charge distribution, and thus the increased polarizability of the atom.

- Molecular orientation with respect to an electric field can affect polarizibility (labeled Orientation-dependent), except for molecules that are: tetrahedral, octahedral or icosahedral (labeled Orientation-independent). This factor is more important for unsaturated molecules that contain areas of electron-dense regions, such as 2,4-hexadiene. Greatest polarizability in these molecules is achieved when the electric field is applied parallel to the molecule rather than perpendicular to the molecule.

2.4.4 Polarizability Influences Dispersion Forces

The dispersion force is the weakest intermolecular force. It is an attractive force that arises from surrounding temporary dipole moments in nonpolar molecules or species. These temporary dipole moments arise when there are instantaneous deviations in the electron clouds of the nonpolar species. Surrounding molecules are influenced by these temporary dipole moments and a sort of chain reaction results in which subsequent weak, dipole-induced dipole interactions are created. These cumulative dipole- induced dipole interactions create attractive dispersion forces. Dispersion forces are the forces that make nonpolar substances condense to liquids and freeze into solids when the temperature is low enough.

Polarizability affects dispersion forces in the following ways:

- As polarizability increases, the dispersion forces also become stronger. Thus, molecules attract one another more strongly and melting and boiling points of covalent substances increase with larger molecular mass.

- Polarazibility also affects dispersion forces through the molecular shape of the affected molecules. Elongated molecules have electrons that are easily moved increasing their polarizability and thus strengthening the dispersion forces, In contrast, small, compact, symmetrical molecules are less polarizable resulting in weaker dispersion forces.

The relationship between polarizability and dispersion forces can be seen in the following equation, which can be used to quantify the interaction between two like nonpolar atoms or molecules:

$$V = -3\alpha 2I/4r6 \quad (2)$$

where

- r is the distance between the atoms or molecules,
- I is the first ionization energy of the atom or molecule, and
- α is the polarizability constant expressed in units of m^3.

2.6 FAZAN'S RULE

Fajans' rule predicts whether a chemical bond will be covalent or ionic. Few ionic bonds have partial covalent characteristics which were first discussed by Kazimierz Fajans in 1923. In the time with the help of X-ray crystallography, he was able to predict ionic or covalent bonding with the attributes like ionic and atomic radius.

We classify certain compounds as ionic and other compounds as covalent. Now if we were to ask the question, amongst the alkali chlorides, which is the most ionic? To answer these kinds of questions, we employ Fajans' rules as a tool.

Fajan's Rules

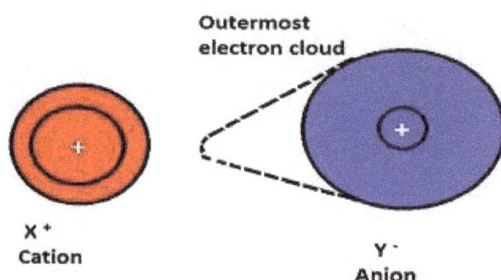

To understand the Fajans' rule, Let us first understand a few terms:

2.5.1 Polarising power

It is the extent to which a cation can polarise an anion. It is proportional to charge density. Charge density is the ratio of charge to volume. Polarising power α Charge density. More the charge density, greater is the polarising power for that cation.

2.5.2 Polarisability

It is the extent to which an ion can be polarised. It can also be called as the ease with which an ion can be polarised. Polarisation is the distortion of a spherically symmetric electron cloud to an unsymmetric cloud.

2.5.3 Postulates of Fajans' Rule

The rule can be stated on the basis of 3 factors, which are:

- ◉ Size of the ion: Smaller the size of cation, the larger the size of the anion, greater is the covalent character of the ionic bond.

- ◉ The charge of Cation: Greater the charge of cation, greater is the covalent character of the ionic bond.

- ◉ Electronic configuration: For cations with same charge and size, the one, with (n-1)dn nso which is found in transition elements have greater covalent character than the cation with ns2 np6 electronic configuration, which is commonly found in alkali or alkaline earth metals.

2.5.4 Explanation of Fazans' Rule

Rule 1: The first rule speaks about the polarising power of the cation. If the cation is smaller, then we can say that the volume of the ion is less. If the volume is less, we can conclude that the charge density of the ion would be high.

Since the charge density is high, the polarising power of the ion would be high. This makes the compound to be more covalent.

Rule 2: The second rule speaks about the polarizability of the anion. Larger the anion, less is the effective nuclear charge that holds the valence electron of the ion in place. Since the last electron is loosely bound in large anions, it can easily be polarised by a cation, thereby making the compound more covalent.

Rule 3: The third rule is a special case. Let us use an example to explain this point.

Example: If we want to find the more covalent compound among $HgCl_2$ and Calcium Chloride we cannot use size as a factor to conclude. This is because both Hg^{2+} and Ca^{2+} are of almost equal size. To explain this, we employ the third rule.

The electronic configuration of Hg^{2+} is $6s^0\ 5d^{10}$. This configuration is called pseudo-octet because d orbital is fully filled, but the element does not have 8 electrons or an octet.

We know that d orbitals are not good at shielding, so we can say that the anion (Cl^-) would be more polarised because the d orbital is poor at shielding making $HgCl_2$ more covalent than $CaCl_2$ because Ca^{2+} ion has a noble gas configuration.

Now to answer the question that we asked first, amongst the alkali chlorides, which one is the most covalent?

Since the anion is the same, we have to compare the cations. According to Fajans' rules, smaller the cation, more is the covalency. Therefore, LiCl is the most covalent.

Let us Understand Fajans' Rule Using a Detailed Illustration:

Consider Aluminum Iodide (AlI3)

This is an ionic bond which was formed by transfer of electrons.

- ⦿ The iodine being bigger in size has a lesser effective nuclear charge. Thus, the bonding electrons are attracted lesser towards the Iodine nucleus.

- ⦿ On the contrary, the aluminium having three positive charges attracts the shared pair of electron towards itself.

- ⦿ This leads to insufficient charge separation for it to be ionic and so it results in the development of covalent character in AlI_3.

Consider Aluminium Fluoride (AlF3)

This is an ionic bond which was also formed by transfer of electron. But here the fluorine being smaller in size attracts the shared pair of an electron more towards itself and so there is sufficient charge separation to make it ionic.

Examples on Fajans' rule

Illustration 1: Which compound should theoretically the most ionic and the most covalent amongst the metal halides?

Solution: The smallest metal ion and the largest anion should technically be the most covalent Therefore, LiI is the most covalent.

The largest cation and the smallest anion should be the most ionic. Therefore, CsF should be the most ionic.

Illustration 2: Arrange the following according to the increasing order of covalency:

- ⊙ NaF, NaCl, NaBr, NaI
- ⊙ LiF, NaF, KF, RbF, CsF

Solution:

1. Since the cation is the same, compare the anions. Amongst the anions, larger the size more would be the covalency. Therefore the order is: NaF < NaCl < NaBr < NaI

2. Here the anion is the same, so we compare with cations. Smaller the cation more is the covalency.

Therefore the order is: CsF < RbF < KF < NaF < LiF

Illustration 3: The melting point of KCl is higher than that of AgCl though the crystal radii of Ag^+ and K^+ ions are almost the same.

Solution : Now whenever any comparison is asked about the melting point of the compounds which are fully ionic from the electron transfer concept it means that the compound having lower melting point has got lesser amount of ionic character than the other one. To analyse such a question first find out the difference between the 2 given compounds. Here in both the compounds the anion is the same. So the deciding factor would be the cation. Now if the anion is different, then the answer should be from the variation of the anion. Now in the above example, the difference of the cation is their electronic configuration. $K^+ = [Ar]$; $Ag^+ = [Kr] 4d^{10}$. This is now a comparison between a noble gas core and pseudo noble gas core, the analysis of which we have already done. So, try to finish off this answer.

Percentage of Ionic Character

Every ionic compound having some percentage of covalent character according to Fajan's rule. The percentage of ionic character in a compound having some covalent character can be calculated by the following equation.

The percent ionic character = Observed dipole moment/Calculated dipole moment assuming 100% ionic bond × 100

Example: Dipole moment of KCl is 3.336×10^{-29} coulomb metre which indicates that it is highly polar molecule. The interatomic distance between K^+ and Cl^- is 2.6×10^{-10} m. Calculate the dipole moment of KCl molecule if there were opposite charges of one fundamental unit located at each nucleus. Calculate the percentage ionic character of KCl.

Solution: Dipole moment μ = e × d coulomb metre

For KCl d = 2.6×10^{-10} m

For complete separation of unit charge

e = 1.602×10^{-19} C

Hence μ = $1.602 \times 10^{-19} \times 2.6 \times 10^{-10}$ = 4.1652×10^{-29} Cm

μKCl = 3.336×10^{-29} Cm

∴ % ionic character of KCl = $3.336 \times 10^{-29} / 4.165 \times 10^{-29}$ = 80.09%

Example. Calculate the % of ionic character of a bond having length = 0.83 Å and 1.82 D as it's observed dipole moment.

Solution: To calculate μ considering 100% ionic bond

= $4.8 \times 10^{-10} \times 0.83 \times 10^{-8}$ esu cm

= $4.8 \times 0.83 \times 10^{-18}$ esu cm = 3.984 D

∴ % ionic character = 1.82/3.984 × 100 = 45.68

The example given above is of a very familiar compound called HF. The % ionic character is nearly 43.25%, so the % covalent character is (100 – 43.25) = 56.75%. But from the octet rule HF should have been a purely covalent compound but actually it has some amount of ionic character in it, which is due to the electronegativity difference of H and F. Similarly knowing the bond length and observed dipole moment of HCl, the % ionic character can be known. It was found that HCl has 17% ionic character. Thus it can be clearly seen that although we call HCl and HF as covalent compounds but it has got appreciable amount of ionic character. So from now onwards we should call a compound having more of ionic less of covalent and vice versa rather than fully ionic or covalent.

03 | COVALENT BONDING

3.1 INTRODUCTION: AN OVERVIEW

Covalent bonding stands as one of the fundamental pillars of chemical bonding, offering a window into the intricate dance of electrons that defines the behaviour of molecules. Unlike ionic bonding, where electrons are transferred between atoms, covalent bonding involves the sharing of electrons between atoms to achieve a stable electron configuration. This sharing of electrons forms the backbone of countless molecules, from the simplest diatomic gases to the most complex organic compounds that populate the natural world.

At its essence, covalent bonding embodies a delicate equilibrium between attraction and repulsion, as atoms strive to attain the elusive state of electron stability. Through the sharing of electrons, atoms forge intimate connections, binding together to form molecules with unique structures, properties, and functionalities. This sharing of electrons enables atoms to satisfy the octet rule, a guiding principle in covalent bonding that dictates atoms tend to achieve a full outer electron shell, typically composed of eight electrons.

The allure of covalent bonding lies not only in its simplicity but also in its versatility, enabling the creation of an astonishing array of molecular architectures and chemical compounds. From the intricate networks of carbon-carbon bonds that underpin the diversity of organic molecules to the polar covalent bonds that give rise to the enchanting properties of water, covalent bonding permeates every facet of chemistry and the natural world.

In this exploration of covalent bonding, we embark on a journey to unravel its mysteries, delving into the mechanisms that govern the sharing of electrons and the factors that influence the strength and nature of covalent bonds. From the seminal insights of early pioneers to the cutting-edge techniques driving contemporary research, we traverse the intricate landscape of chemical bonding, seeking to uncover the fundamental principles that govern the behaviour of atoms and molecules.

Through experimentation, observation, and theoretical abstraction, we endeavour to illuminate the profound significance of covalent bonding in shaping the world around us, from the molecules that sustain life to the materials that propel technological innovation. Join us as we delve into the captivating realm of covalent bonding, where electrons forge connections that transcend the boundaries of individual atoms, giving rise to the rich tapestry of matter that surrounds us.

3.2 ELECTRONIC CONFIGURATION

The outermost electrons surrounding an atom (the **valence electrons**) are responsible for the number and type of bonds that a given atom can form with other atoms, and are responsible for the chemistry of the atom. The shape of the modern periodic table reflects the arrangement of electrons by grouping elements together into s, p, d, and f blocks. If we compare the arrangement of electrons, the logical grouping of elements becomes apparent. The **electron configuration** (the arrangement of electrons in an element) is a direct result of the work of Bohr, Heisenberg, Schroedinger, and other physicists in the early 20th century.

Elements are arranged in horizontal rows, called **periods**. Each period is a **principal energy level**, numbered 1, 2, 3, 4, etc. The first principal energy level starts with hydrogen, the second level starts with lithium, the third with sodium, etc. Each principal energy level holds a maximum number of electrons equal to 2n2, where n is the principal energy level number. The second energy level starting with lithium holds $2(2^2)$ = 8 electrons maximum; the third level starting with sodium holds $2(3^2)$ = 18; the fourth can hold $2(4^2)$ = 32, and so on.

Principal energy levels are divided into **sublevels** following a distinctive pattern, shown in Table 1 below.

Table 1. Principal energy levels and their sublevels.

Principal level 1	Sublevels
2	s
3	s, p
4	s, p, d s, p, d, f
5	s, p, d, f, g
6	s, p, d, f, g, h
7	s, p, d, f, g, h, i

The principal energy levels and sublevels build on top of each other. The sodium atom in the third principal energy level has the first principal level composed of an s--sublevel, and the second principal level composed of s-- and p-- sublevels, underneath the third principal level. Probably the simplest mental image would be an onion. Each principal level corresponds to a layer of the onion. As you move farther from the center of the onion, the layers get larger. Larger layers can be sub--divided into more pieces (sublevels) than can smaller layers.

Each sublevel is in turn divided into *orbitals*, specific locations for the electrons. The number of orbitals for each sublevel also follows a distinctive pattern, shown in Table 2 below.

Table 2. Total orbitals for each type of sublevel.

Sublevel	Total orbitals
s	1
p	3
d	5
f	7
g	9
h	11
i	13

All s--sublevels have 1 orbital, regardless of whether they are 1s, or 2s, or 3s, etc. All p--sublevels contain 3 orbitals, and d--sublevels contain 5 orbitals, and so on. Every orbital, regardless of the sublevel, holds a maximum of 2 electrons.

In order to compare the arrangement of electrons around atoms, we need some way of predicting the filling order of the principal levels and sublevels. Figure 1 below guides us in arranging the electrons.

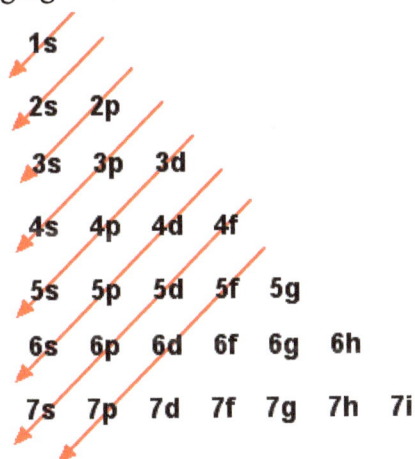

Figure 1. Filling order of principal levels and sublevels.

Using Figure 1 is as simple as following the arrows. Start with the top arrow, and follow its direction from tail to head; you see that the first arrow passes through 1s, and the 1s sublevel is the first to fill with electrons. When the 1s sublevel is full, continue to the second arrow which passes through the 2s sublevel. The third arrow passes through the 2p and 3s sublevels (in this order), while the fourth arrow passes through the 3p and 4s sublevels, in order. The fifth arrow passes through the 3d, 4p, and 5s sublevels in order.

Using the order provided from Figure 1, and remembering the total number of orbitals in each sublevel, we can write electronic configurations for the elements. The **electronic configuration** is the electronic structure of the atoms; the specific levels, sublevels, and number of electrons occupying orbitals for a given atom.

Let's look at group 1A elements. Hydrogen has 1 electron, and the electronic configuration is $1s1$. This tells us that the first principal energy level is being filled, the s--sublevel is the specific location within the principal energy level, and that there is 1 electron in the s--orbital (the superscript "1" is the number of electrons).

Electronic configurations for all group 1A elements are given below.

H (1 electron):	1s1
Li (3 electrons):	1s2, 2s1
Na (11 electrons):	1s2, 2s2, 2p6, 3s1
K (19 electrons):	1s2, 2s2, 2p6, 3s2, 3p6, 4s1
Rb (37 electrons):	1s2, 2s2, 2p6, 3s2, 3p6, 4s2, 3d10, 4p6, 5s1
Cs (55 electrons):	1s2, 2s2, 2p6, 3s2, 3p6, 4s2, 3d10, 4p6, 5s2, 4d10, 5p6, 6s1
Fr (87 electrons):	1s2, 2s2, 2p6, 3s2, 3p6, 4s2, 3d10, 4p6, 5s2, 4d10, 5p6, 6s2, 4f14, 5d10, 6p6, 7s1

Notice that in all cases the outermost energy level contains a single sublevel: an s type sublevel, and this sublevel contains 1 electron. The outside layers of these elements are very similar. The group 2A elements contain one additional electron, and in all cases this additional electron is in the s type sublevel. Beryllium will be $1s2, 2s2$. Magnesium will resemble sodium with an additional electron ($1s2, 2s2, 2p6, 3s2$), calcium will resemble potassium with an additional electron, and so on.

This pattern follows throughout the periodic table. The group 3A elements all have an outer layer following the pattern $ns2, np1$ (where n is the principal energy level number). Boron's outside layer is $2s2, 2p1$, aluminum's is $3s2, 3p1$, and so on.

Group 4A elements follow the pattern $ns2, np2$; Group 5A's pattern is $ns2, np3$, and so on.

Group 8A, the Noble gases, have a very interesting electronic structure.

He (2 electron):	1s2
Ne (10 electrons):	1s2, **2s2, 2p6**
Ar (18 electrons):	1s2, 2s2, 2p6, **3s2, 3p6**
Kr (36 electrons):	1s2, 2s2, 2p6, 3s2, 3p6, **4s2**, 3d10, **4p6**
Xe (54 electrons):	1s2, 2s2, 2p6, 3s2, 3p6, 4s2, 3d10, 4p6, **5s2**, 4d10, **5p6**
Rn (86 electrons):	1s2, 2s2, 2p6, 3s2, 3p6, 4s2, 3d10, 4p6, 5s2, 4d10, 5p6, **6s2**, 4f14, 5d10, **6p6**

The outside layer of electrons is full! With the exception of helium, which is too small to hold 8 electrons, all other elements in this group follow the pattern ns2, np6. This arrangement is called a *Noble gas configuration*, 8 electrons completely filling the ns and np sublevels, and this arrangement confers extraordinary stability on these compounds. The Noble gases don't participate in very many chemical reactions, and then only under extreme conditions.

When we get into the *d* block (transition elements) and the *f* block (lanthanides/actinides) the same general pattern holds. Group 3B elements have similar electron configurations, as do group 4B, 5B, and so on.

3.3 LEWIS DOT STRUCTURES

Metallic elements, to the left of the staircase dividing line, tend to loose one or more electrons and form ions. Hydrogen is the single general exception to this trend; when combined with non--metals, hydrogen tends to form covalent compounds.

All of the non--metals have filled s--sublevels, and partially filled p--sublevels. There are a fixed number of electrons on the outside layer of these elements, and this number is equal to the group number. Boron is the only group 3A non--metal, and has 3 electrons in its outside layer. Carbon and silicon are group 4A non--metals and have 4 electrons in their outside layers. Groups 5, 6, 7, and 8A follow similarly.

In 1916, American physical chemist Gilbert Newton Lewis (1875 – 1946) discovered the covalent bond and developed a method of representing bonding between non--metals using simple diagrams called Lewis dot structures. Lewis dot structures are the elements chemical symbol, with the valence electrons arranged uniformly around the four sides of the symbol. Lewis dot structures for the first few non--metals are shown in Figure 2.

$$\cdot \overset{\displaystyle \cdot}{B} \cdot \quad \cdot \overset{\displaystyle \cdot}{C} \cdot \quad \cdot \overset{\displaystyle \cdot \cdot}{\underset{\displaystyle \cdot}{N}} \cdot \quad \overset{\displaystyle \cdot \cdot}{\underset{\displaystyle \cdot \cdot}{:O}} \cdot \quad \overset{\displaystyle \cdot \cdot}{\underset{\displaystyle \cdot \cdot}{:F}} \cdot \quad \overset{\displaystyle \cdot \cdot}{\underset{\displaystyle \cdot \cdot}{:Ne:}}$$

Figure 2. **Lewis dot structures for the first few non--metals.**

Silicon is similar to carbon, with "Si" replacing "C", phosphorous and arsenic are similar to nitrogen; sulfur, selenium, tellurium are similar to oxygen; chlorine, bromine, and iodine are similar to fluorine; and the Noble gases are similar to neon.

In Lewis dot structures, some dots are single, representing single electrons, while others are double representing pairs of electrons. We can predict the number of bonds each element forms by counting the number of single electrons in the Lewis dot structure. Boron can form 3 bonds, carbon can form 4 bonds, nitrogen can form 3, oxygen can form 2 bonds, fluorine forms 1 bond, and neon doesn't form any bonds because it has no single electrons, only pairs.

When one element shares a single electron with another element having a single electron, the shared pair of electrons is a chemical bond. For example, hydrogen (having only 1 electron) can share this electron with fluorine to make hydrogen fluoride (HF). The Lewis dot structure of this compound is:

$$\text{H} \; \overset{\cdot\cdot}{\underset{\cdot\cdot}{\text{F}}} \;$$

Figure3. Lewis dot structure of hydrogen fluoride.

For clarity, the electron from hydrogen is colored red. The chemical bond is composed of one (black) electron from fluorine and one (red) electron from hydrogen.

In water, two hydrogen atoms are connected to an oxygen atom, and the Lewis dot structure of this compound is:

$$\overset{\cdot\cdot}{\underset{\cdot\cdot}{\text{O}}} \; \text{H}$$
$$\text{H}$$

Figure 4. Lewis dot structure of water.

Before anyone gets wrong ideas; the electrons aren't permanently fixed to one particular side of the chemical symbol. We can always distribute the electrons in any fashion that is convenient or pleasing to us, provided that we keep them either as single electrons or as pairs of electrons (no groups of 3 or more electrons).

We could always draw the water molecule as:

$$\text{H} \; \overset{\cdot\cdot}{\underset{\cdot\cdot}{\text{O}}} \; \text{H}$$

Figure 5. Alternate Lewis dot structure of water.

The shared pair will always be somewhere between the two bonded atoms. Often, we show a shared pair as a single line, representing the chemical bond. Using this method we can draw our water molecule as:

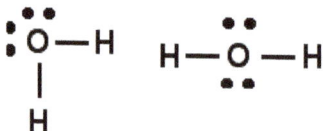

$$\overset{\cdot\cdot}{\underset{\cdot\cdot}{\text{O}}}\!-\!\text{H} \qquad \text{H}\!-\!\overset{\cdot\cdot}{\underset{\cdot\cdot}{\text{O}}}\!-\!\text{H}$$
$$|$$
$$\text{H}$$

Figure 6. Lewis structures of water, showing two equivalent structures.

Strictly speaking, using a combination of lines for bonds and dots for unshared pairs of electrons are *Lewis structures*, not Lewis dot structures. The distinction is not particularly important for this course and you can use either. However, don't try to use both! Some students will draw a single line, and then include two dots as electrons (above the line, below the line, or straddling the line). This is wrong and confusing.

With two exceptions (hydrogen and boron) all non--metals must have a total of 8 electrons. These electrons can be shared pairs (bonds) or unshared pairs.

Hydrogen forms one bond, and is satisfied having 2 electrons. Boron has 3 electrons, and can make 3 bonds for a total of 6 electrons. All other non--metals need 8 electrons.

The structures of many simple molecules can be determined by comparing the number of bonds that each element can form. However, not all molecules are equally simple, and small molecules can sometimes be deceptive. Consider carbon monoxide (CO). Carbon has 4 unshared electrons, and can make 4 bonds. Oxygen has 2 unshared electrons, and can make 2 bonds. What does carbon monoxide look like?

When the simple inspection method doesn't work, we use the "pooled electron" method. We add together the total number of valence electrons for our elements, and then distribute them so that each element has 8 electrons (either as shared pairs or as unshared pairs). For carbon monoxide, carbon contributes 4 electrons, oxygen contributes 6, and the total valence electrons are 10. These 10 electrons have to be distributed so that bonds are formed, and so that both carbon and oxygen have 8 electrons. These two requirements result in the following Lewis dot structure:

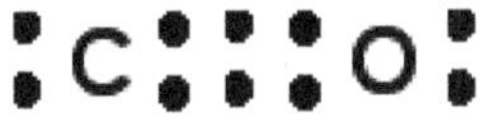

Figure 7. Lewis dot structure of carbon monoxide.

Or if we prefer to use lines for bonds:

Figure 8. Lewis structure of carbon monoxide.

This method works especially well for polyatomic ions. In sulfate (SO_4^{-2}), sulfur contributes 6 electrons, and each oxygen atom contributes 6 electrons. There are two additional electrons indicated by the --2 charge, for a total of 32 electrons.

One possible structure would be a central sulfur atom surrounded by the oxygen atoms:

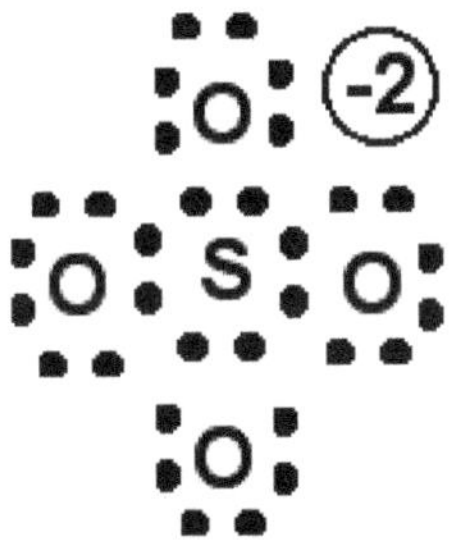

Figure 9. One possible Lewis dot structure for sulfate.

Other arrangements meeting the requirements that all electrons are used and each atom has 8 electrons are possible.

For positively charged polyatomic ions like ammonium and hydronium, we must subtract the positive charge from the valence electrons, since loss of electron(s) results in positive charge(s). Ammonium has a total of 8 valence electrons, as does hydronium, and their Lewis dot structures are:

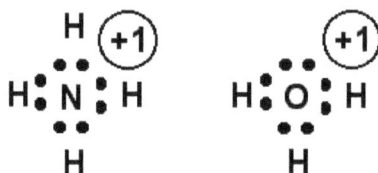

Figure 10. Lewis dot structures of ammonium (left) and hydronium (right).

Sometimes, a compound or ion can have more than one valid structure. An example is carbonate, which has three equivalent Lewis dot structures:

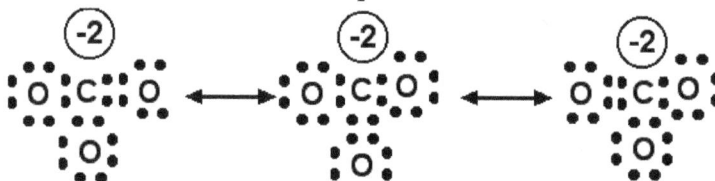

Figure 11. Three Lewis dot structures for the carbonate anion. The three structures are equivalent.

Notice that one of the oxygen atoms shares 4 electrons with carbon (it is double bonded to carbon). In each structure a different oxygen atom is double bonded to carbon. This is generally explained by the idea of **resonance**. Resonance does **NOT** mean that the ion is rapidly interchanging between the three different forms, nor does it mean that there is an equal mixture of the three forms. Instead, the extra carbon--oxygen bond is somehow "spread out" over the entire molecule.

Every carbon--oxygen bond is actually 1.3333 bonds.

Now, this idea that we can have "fractional" bonds doesn't make very much sense if we are talking about bonds being equivalent to two electrons, because what sense is there is talking about a fraction of an electron? However, you need to remember that the Bohr model of the atom, with electrons as small negatively charged particles orbiting the nucleus, is **WRONG** (although very convenient for many purposes). In reality, the electron (when "orbiting" the nucleus) is an electromagnetic wave. Just like two sound waves can combine to form a new sound wave, or two water waves can combine to form a larger water wave, two electromagnetic waves can combine to form a new electromagnetic wave. When two "electron--waves" combine, they form a new "electron--wave" that we call a "chemical bond". Having 1/3 of an electron is a difficult notion, but having one wave 1/3 as large as another wave isn't nearly as difficult to picture. Imagining how two electrons, as negatively charged particles, can hold two atoms together is difficult.

Imagining two electrons combining to form an "electron--wave" may not help very much, but it is a better description of the chemical bond.

I'm going to continue talking about pairs of electrons as if they are actual small negatively charged bits of matter, but when they are IN the atom, they aren't.

3.3.1 VSEPR theory

Valence shell electron pair repulsion theory (VSEPR) allows us to predict the geometry of a molecule, based on the arrangement of atoms and unshared electron pairs around a central atom. As we have seen from our Lewis dot structures, each atom tends to have four pair of electrons around it. If a molecule is sufficiently large (5 atoms total), then we have 4 atoms arranged around fifth atom at the center.

Methane (CH_4) is a good example of this arrangement.

When we draw a methane molecule on a flat surface, the flat surface forces 2--dimensional geometry (unless we try to make some sort of perspective drawing).

Below are some common representations of a methane molecule.

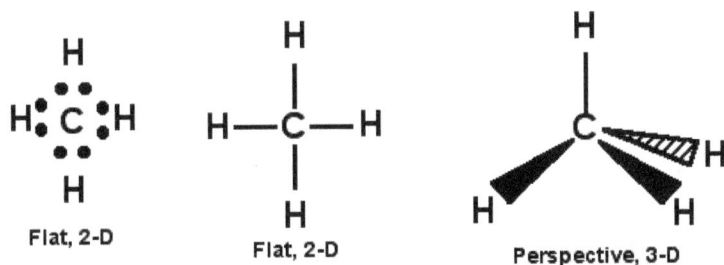

Flat, 2-D   Flat, 2-D   Perspective, 3-D

Figure 12. Three different representations of the methane molecule.

The perspective drawing tries to emphasize the 3--dimensional nature of molecules. By convention, the dark triangular lines are bonds projecting out of the surface of the paper, while the hashed triangle is a bond projecting into and through the paper. Methane is a tetrahedral molecule – a pyramid made of 4 equal triangles (as opposed to a normal pyramid made of 4 equal triangles and a square base).

Tetrahedral geometry is the only 3--D geometry allowing four objects to be equally distributed around a central fifth object.

VSEPR theory states that a pair of electrons occupies almost the same volume of space as a hydrogen atom. Hydrogen fluoride, water, sulfate, ammonia, and hydronium ion have shapes similar to methane, but with unshared electron pairs replacing hydrogen atoms in some cases. These substances are shown below.

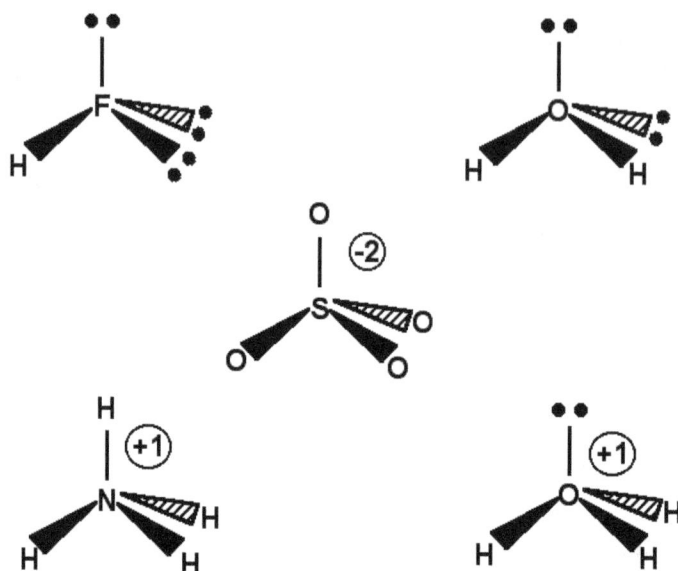

Figure 13. Geometry of hydrogen fluoride, water, sulfate, ammonium, and hydronium ions.

If there are fewer than 4 atoms bonded around a central atom, or if there are no unshared electron pairs, the molecular geometry changes. Table 3 summarizes various possible combinations of atoms and unshared pairs, and the resulting geometry. These "rules" apply to central atoms obeying the octet rule. Some elements, such as sulfur, have the ability for form "expanded octets", resulting in more than 8 electrons around sulfur. These are exceptions, and aren't particularly useful in learning the general pattern exhibited by most compounds, so we won't worry about these examples here.

Table 3. Molecular geometry based on bonded atoms and lone pairs.

# of atoms bonded to central atom	# of unshared electron pairs	Molecular geometry	Examples
2	0	Linear	Carbon dioxide
2	1	Bent	Sulfur dioxide
2	2	Bent	Water
3	0	Trigonal planar	Boron trifluoride
3	1	Trigonal pyramidal	Ammonia
4	0	Tetrahedral	Methane

3.3.2 Bond Polarity and Molecular Polarity

When identical atoms share electrons in a chemical bond, the sharing must be exactly equal. Consider two hydrogen atoms bonded to form a molecule. Neither hydrogen atom can exert greater control of the electrons than the other. The electron pair is shared equally between the two atoms, because the two atoms are identical.

When dissimilar atoms share electrons, the situation is different. Each element has a characteristic *electronegativity*, a chemical property describing the tendency of an atom to attract electrons towards itself. Table 4 shows Pauling electronegativity values for selected non--metal compounds. "Pauling electronegativities" are named for American chemist Linus Carl Pauling (1901 – 1994), who developed the concept of electronegativity. Pauling was one of the founders of quantum chemistry and of molecular biology. He is the only person to win two unshared Nobel Prizes, the first for Chemistry and the second for Peace.

We can estimate how equally electrons are shared between elements by comparing electronegativities for pairs of elements – simply subtract the smaller value from the larger. This result, the difference in electronegativity (DEN) is used to determine the *bond polarity*.

If DEN < 0.6, then the bond is considered to be "nonpolar covalent". It is essentially the same as a bond between identical atoms. Carbon--hydrogen bonds have DEN = 0.4, so they are nonpolar covalent.

Table 4. Pauling electronegativity values for selected non--metals.

H	B	C	N	O	F
2.1	2.0	2.5	3.0	3.4	4.0
		Si 1.9	P 2.2	S 2.6	Cl 3.2
				Se 2.5	Br 3.0
					I 2.7

If DEN < 1.6, then the bond is considered to be "polar covalent". In this bond, the electrons are shared, but they are strongly attracted to the element having the higher electronegativity value. Hydrogen--oxygen bonds have DEN = 1.3, and the shared electrons are attracted towards oxygen and away from hydrogen.

If DEN > 2.0, the bond is considered ionic. The only example we have of this with the non--metals is the silicon--fluorine bond (DEN = 2.1). However, many metals (not shown) have very low electronegativity values and readily form ionic bonds.

Sodium or calcium's electronegativity is 1.0. Sodium--oxygen bonds have DEN = 2.4, sodium--chlorine bonds have DEN = 2.2. Generally, we classify any compound between metals and non--metals as ionic as a matter of course.

If DEN is between 1.6 and 2.0, then the bond classification depends on the type of elements combined. If one of the elements is a metal, then the bond is considered ionic, while if both elements are non--metals, then the bond is polar covalent. Hydrogen fluoride has DEN = 1.9, but since both elements are non--metals the bond is polar covalent. Scandium's electronegativity is 1.3, and scandium chloride has DEN = 1.9. Since scandium is a metal, this bond is ionic.

We can readily show the attraction of shared pairs towards one element by using arrows instead of straight lines, with the arrowhead pointing towards the element that attracts the electrons (Figure 14). If the bond is nonpolar covalent, we can use either a line or a double--headed arrow. Notice the "δ+" and

"δ--" symbols – these represent partial + and – charges, due to the imbalance in electron sharing. In water for example, the electrons are strongly attracted to oxygen, resulting in oxygen being a little bit negative. The hydrogen atoms are slightly deprived of electrons and become a little bit positive.

Molecular polarity is the result of unshared electron pairs, bond polarity, and geometry. A polar molecule will have one side or end of the molecule slightly positive and the opposite end or side slightly negative. If you use the following guidelines, you will rarely go wrong in assigning molecular polarity. Once again, these guidelines apply only to central atoms obeying the octet rule. Atoms with expanded octets have modified guidelines.

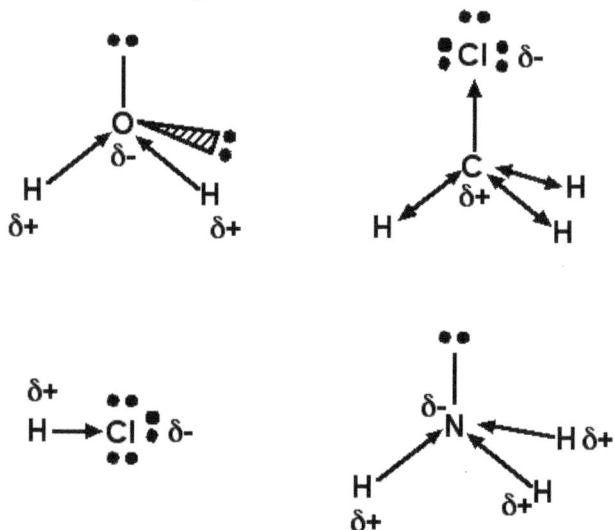

Figure 14. Bond polarities are shown using arrows. Shared electrons are drawn towards the element having higher electronegativity.

1. Does the central atom have unshared electron pairs? If "yes", then the molecule is polar. Water and ammonia are classic examples; unshared electrons on the central atom guarantee that the molecule is polar.
2. If there are no unshared electron pairs on the central atom, then does the atom have polar covalent or ionic bonds? If all of the bonds are nonpolar covalent, then the molecule is nonpolar. Methane (CH4) is the classic example of this guideline in action. The Lewis dot structure shows that the central

carbon atom does not have any unshared electron pairs. The carbon-- hydrogen bonds have DEN = 0.4, clearly nonpolar covalent bonds. The methane molecule is therefore nonpolar.

3. If there are no unshared electron pairs on the central atom, but there are polar covalent bonds, then the molecule's geometry determines molecular polarity. Geometry can cancel the effects of bond polarity. The simplest way to see this effect is to consider a series of compounds, CH4, CH3F, CH2F2, CHF3, and CF4 (Figure 15). Methane is nonpolar as described above. Carbon-- fluorine bonds are polar covalent, shared electrons are attracted to the fluorine atom(s), and the molecule has slightly positive/negative sides. In CF4, the "outside" of the molecule is uniformly negative, while the "inside" (the carbon atom) is slightly positive. However, nothing can get near the slightly positive carbon atom, so the molecule is effectively non--polar.

4. If the substance is made from ions, then it is polar.

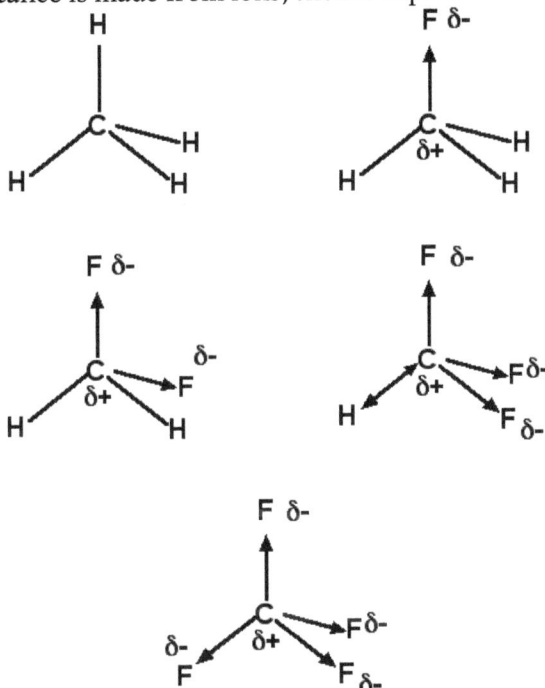

Figure 15. Bond polarity effects can be cancelled out by the molecules geometry.

Molecular polarity explains a variety of chemical and physical properties. One example is the expression "like dissolves like". Polar and ionic substances readily dissolve in polar materials, and are insoluble in nonpolar substances.

Sodium chloride (ionic) or ammonia (polar) dissolves readily in water (polar).

Hexane (C6H14, no unshared pairs and nonpolar bonds) is nonpolar and doesn't dissolve particularly well in water. **Hydrogen bonding** is an intermolecular force in which the hydrogen atom attached to an oxygen, nitrogen, or fluorine atom is attracted to the oxygen, nitrogen, or fluorine atom of an adjacent molecule. Water is the classic example of hydrogen bonding, either with another water molecule or with any molecule containing oxygen, nitrogen, or fluorine (Figure 16).

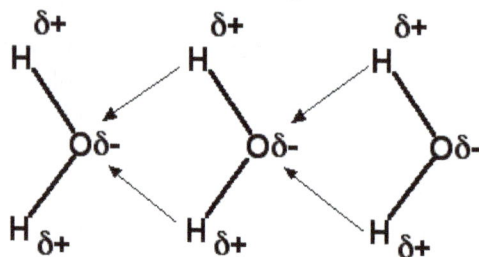

Hydrogen bonding in water.

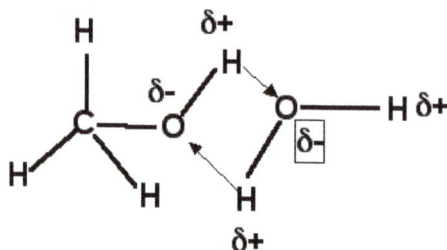

Hydrogen bonding between water and alcohol.

Hydrogen bonds are shown as arrows.

Figure 16. Hydrogen bonding examples.

◉ **Example: Carbon Dioxide, CO_2**

1.   Carbon is the least electronegative atom, hence the central.
2.   Carbon has 4 valence electrons, each oxygen atom has 6.

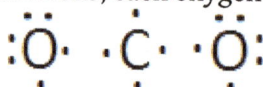

3.   Single Bonded.

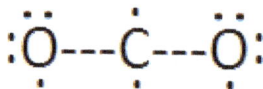

4.   Double bonded to remove lone electrons.

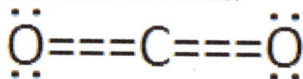

5.   Total valence electrons = 4+ 2(6) = **16** electrons

2 double bonds = 2 x 4 electrons = 8 electrons

4 lone pairs = 2 x 4 electrons = 8 electrons

8 + 8 = **16** electrons

3.4 RESONANCE AND FORMAL CHARGE

Resonance structures are a set of two or more Lewis Structures that collectively describe the electronic bonding of a single polyatomic species including fractional bonds and fractional charges. Resonance structures are capable of describing delocalized electrons that cannot be expressed by a single Lewis formula with an integral number of covalent bonds.

Sometimes one Lewis Structure is not Enough

Sometimes, even when formal charges are considered, the bonding in some molecules or ions cannot be described by a single Lewis structure. Resonance is a way of describing delocalized electrons within certain molecules or polyatomic ions where the bonding cannot be expressed by a single Lewis formula. A molecule or ion with such delocalized electrons is represented by several contributing structures (also called resonance structures or canonical forms). Such is the case for ozone (O_3), an allotrope of oxygen with a V-shaped structure and an O–O–O angle of 117.5°.

3.4.1 Ozone (O3)

1. We know that ozone has a V-shaped structure, so one O atom is central:

$$O$$
$$O \quad O$$

2. Each O atom has 6 valence electrons, for a total of 18 valence electrons.

3. Assigning one bonding pair of electrons to each oxygen–oxygen bond gives with 14 electrons left over.

4. If we place three lone pairs of electrons on each terminal oxygen, we obtain and have 2 electrons left over.

5. At this point, both terminal oxygen atoms have octets of electrons. We therefore place the last 2 electrons on the central atom:

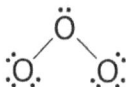

6. The central oxygen has only 6 electrons. We must convert one lone pair on a terminal oxygen atom to a bonding pair of electrons—but which one? Depending on which one we choose, we obtain either

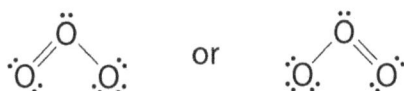

or

Which is correct? In fact, neither is correct. Both predict one O–O single bond and one O=O double bond. As you will learn, if the bonds were of different types (one single and one double, for example), they would have different lengths. It turns out, however, that both O–O bond distances are identical, 127.2 pm, which is shorter than a typical O–O single bond (148 pm) and longer than the O=O double bond in O_2 (120.7 pm).

Equivalent Lewis dot structures, such as those of ozone, are called resonance structures. The position of the atoms is the same in the various resonance structures of a compound, but the position of the electrons is different. Double-headed arrows link the different resonance structures of a compound:

The double-headed arrow indicates that the actual electronic structure is an average of those shown, not that the molecule oscillates between the two structures.

When it is possible to write more than one equivalent resonance structure for a molecule or ion, the actual structure is the average of the resonance structures.

3.4.2 The Carbonate (CO2–3) Ion

Like ozone, the electronic structure of the carbonate ion cannot be described by a single Lewis electron structure. Unlike O_3, though, the actual structure of CO_3^{2-} is an average of *three* resonance structures.

1. Because carbon is the least electronegative element, we place it in the central position:

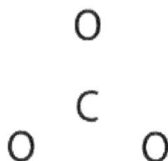

The three oxygens are drawn in the shape of a triangle with the carbon at the center of the triangle.

2. Carbon has 4 valence electrons , each oxygen has 6 valence electrons , and there are 2 more for the –2 charge. This gives $4 + (3 \times 6) + 2 = 24$ valence electrons.

3. Six electrons are used to form three bonding pairs between the oxygen atoms and the carbon:

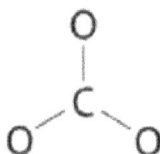

4. We divide the remaining 18 electrons equally among the three oxygen atoms by placing three lone pairs on each and indicating the -2 charge:

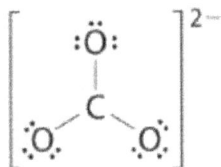

$$\left[\begin{array}{c} \ddot{\text{O}}: \\ | \\ \text{C} \\ :\ddot{\text{O}} \quad \ddot{\text{O}}: \end{array} \right]^{2-}$$

The Lewis dot structure has a central carbon that is bonded to 3 oxygens. Each oxygen has 3 lone pairs. The molecule is inside square brackets and has a charge of minus 2.

5. No electrons are left for the central atom.

6. At this point, the carbon atom has only 6 valence electrons , so we must take one lone pair from an oxygen and use it to form a carbon–oxygen double bond. In this case, however, there are *three* possible choices:

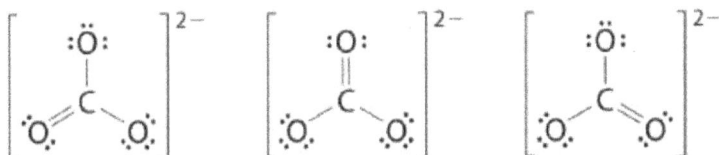

$$\left[\begin{array}{c} :\ddot{\text{O}}: \\ | \\ \text{C} \\ \text{O}\!=\!\!\!\quad\!\!\!:\ddot{\text{O}}: \end{array} \right]^{2-} \left[\begin{array}{c} :\text{O}: \\ \| \\ \text{C} \\ :\ddot{\text{O}} \quad \ddot{\text{O}}: \end{array} \right]^{2-} \left[\begin{array}{c} :\ddot{\text{O}}: \\ | \\ \text{C} \\ :\ddot{\text{O}} \quad \text{O}: \end{array} \right]^{2-}$$

As with ozone , none of these structures describes the bonding exactly. Each predicts one carbon–oxygen double bond and two carbon–oxygen single bonds, but experimentally all C–O bond lengths are identical. We can write resonance structures (in this case, three of them) for the carbonate ion:

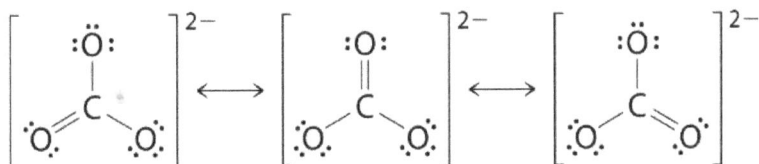

$$\left[\begin{array}{c} :\ddot{\text{O}}: \\ | \\ \text{C} \\ \text{O}\!=\!\!\!\quad\!\!\!:\ddot{\text{O}}: \end{array} \right]^{2-} \longleftrightarrow \left[\begin{array}{c} :\text{O}: \\ \| \\ \text{C} \\ :\ddot{\text{O}} \quad \ddot{\text{O}}: \end{array} \right]^{2-} \longleftrightarrow \left[\begin{array}{c} :\ddot{\text{O}}: \\ | \\ \text{C} \\ :\ddot{\text{O}} \quad \text{O}: \end{array} \right]^{2-}$$

The resonance structure includes all three Lewis dot structures with double headed arrows between them.

The actual structure is an average of these three resonance structures.

3.4.3 The Nitrate (NO–3) ion

1. Count up the valence electrons: $(1*5) + (3*6) + 1(\text{ion}) = \textbf{24}$ electrons

2. Draw the bond connectivities:

$$\begin{array}{c} \text{O} \\ | \\ \text{N} \\ \text{O} \quad\quad \text{O} \end{array}$$

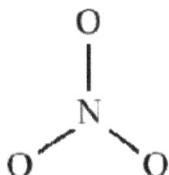

The three oxygens are drawn in the shape of a triangle with the nitrogen at the center of the triangle.

3.   Add octet electrons to the atoms bonded to the center atom:

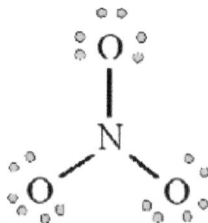

4.   Place any leftover electrons (24-24 = **0**) on the center atom:

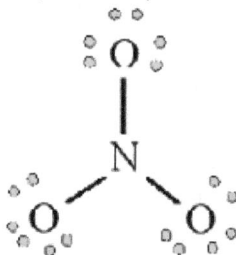

5.   Does the central atom have an octet?

⦿ **NO,** it has 6 electrons

⦿ Add a multiple bond (first try a double bond) to see if the central atom can achieve an octet:

A double bond is added between one oxygen and the central nitrogen. The molecule has a negative charge.

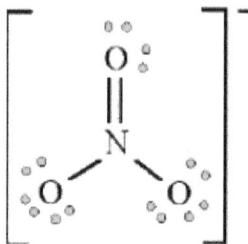

6.   Does the central atom have an octet?

⦿ YES

⦿ Are there possible resonance structures? YES

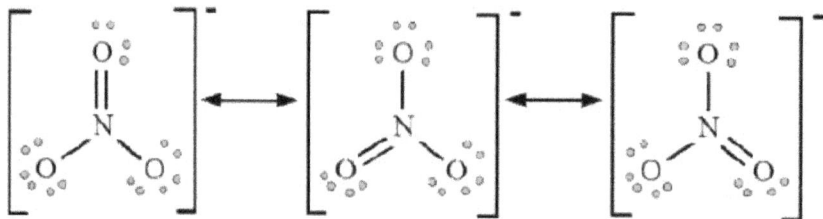

Note: We would expect that the bond lengths in the NO^{-3} ion to be somewhat shorter than a single bond.

Example : Benzene

Benzene is a common organic solvent that was previously used in gasoline; it is no longer used for this purpose, however, because it is now known to be a carcinogen. The benzene molecule (C6H6C6H6) consists of a regular hexagon of carbon atoms, each of which is also bonded to a hydrogen atom. Use resonance structures to describe the bonding in benzene.

Given: molecular formula and molecular geometry

Asked for: resonance structures

Strategy: Draw a structure for benzene illustrating the bonded atoms. Then calculate the number of valence electrons used in this drawing.

A. Subtract this number from the total number of valence electrons in benzene and then locate the remaining electrons such that each atom in the structure reaches an octet.

B. Draw the resonance structures for benzene.

Solution:

A. Each hydrogen atom contributes 1 valence electron, and each carbon atom contributes 4 valence electrons, for a total of (6 × 1) + (6 × 4) = 30 valence electrons If we place a single bonding electron pair between each pair of carbon Atoms and between each carbon and a hydrogen atom, we obtain the following:

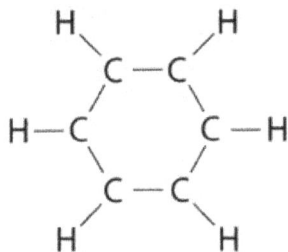

Each carbon atom in this structure has only 6 electrons and has a formal charge of +1, but we have used only 24 of the 30 valence electrons.

B. If the 6 remaining electrons are uniformly distributed pairwise on alternate carbon atoms, we obtain the following:

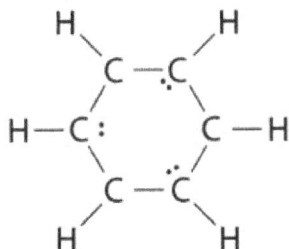

Three carbon atoms now have an octet configuration and a formal charge of –1, while three carbon atoms have only 6 electrons and a formal charge of +1. We can convert each lone pair to a bonding electron pair, which gives each atom an octet of electrons and a formal charge of 0, by making three C=C double bonds.

C. There are, however, two ways to do this:

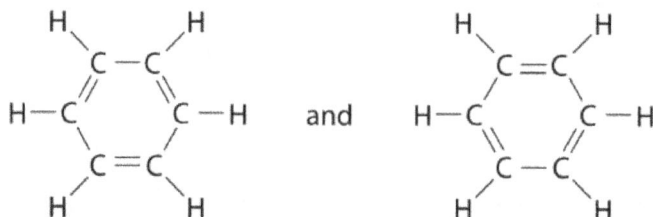

Each structure has alternating double and single bonds, but experimentation shows that each carbon–carbon bond in benzene is identical, with bond lengths (139.9 pm) intermediate between those typically found for a C–C single bond (154 pm) and a C=C double bond (134 pm). We can describe the bonding in benzene using the two resonance structures, but the actual electronic structure is an average of the two. The existence of multiple resonance structures for aromatic hydrocarbons like benzene is often indicated by drawing either a circle or dashed lines inside the hexagon:

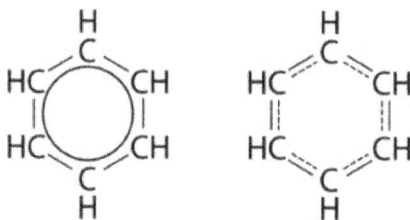

Benzene

The sodium salt of nitrite is used to relieve muscle spasms. Draw two resonance structures for the nitrite ion (NO_2^-).

If several reasonable resonance forms for a molecule exists, the "actual electronic structure" of the molecule will probably be intermediate between all the forms that you can draw. The classic example is benzene in Example 9.8.19.8.1. One would expect the double bonds to be shorter than the single bonds, but if one overlays the two structures, you see that one structure has a single bond where the other structure has a double bond. The best measurements that we can make of benzene do not show two bond lengths - instead, they show that the bond length is intermediate between the two resonance structures.

Resonance structures is a mechanism that allows us to use all of the possible resonance structures to try to predict what the actual form of the molecule would be. Single bonds, double bonds, triple bonds, +1 charges, -1 charges, these are our limitations in explaining the structures, and the true forms can be in between - a carbon-carbon bond could be mostly single bond with a little bit of double bond character and a partial negative charge, for example.

Summary

Some molecules have two or more chemically equivalent Lewis's electron structures, called resonance structures. Resonance is a mental exercise and method within the of bonding that describes the delocalization of electrons within molecules. These structures are written with a double-headed arrow between them, indicating that none of the Lewis structures accurately describes the bonding but that the actual structure is an average of the individual resonance structures. Resonance structures are used when one Lewis structure for a single molecule cannot fully describe the bonding that takes place between neighboring   atoms relative to the empirical data for the actual bond lengths between those atoms. The net sum of valid resonance structures is defined as a resonance hybrid, which represents the overall delocalization oaf electrons within the molecule. A molecule that has several resonance structures is more stable than one with fewer. Some resonance structures are more favorable than others.

3.5 POLAR AND NONPOLAR COVALENT BONDS

3.5.1 INTRODUCTION

Covalent bonds form the backbone of molecular structures, dictating the properties and behaviours of countless substances in the natural world. These bonds arise when atoms share electrons, establishing a stable configuration that satisfies their valence electron requirements. However, not all covalent bonds are created equal. The distinction between polar and nonpolar covalent bonds lies in the distribution of electrons within the bond. In polar covalent bonds, electrons are unevenly shared between atoms due to differences in electronegativity, the tendency of an atom to attract electrons towards itself. This results in a partial negative charge on one atom and a partial positive charge on the other, creating an electric dipole within the molecule. Such polarity influences the molecule's interactions with other substances, its solubility, and its behaviour in various environments. Conversely, nonpolar covalent bonds arise when atoms share electrons equally, typically occurring between atoms with similar electronegativities. In these bonds, there is no significant charge separation within the molecule, leading to symmetric electron distributions. As a result, nonpolar molecules tend to exhibit different properties and behaviours compared to their polar counterparts. Understanding the nature of polar and nonpolar covalent bonds is fundamental to comprehending the diverse array of molecular structures and their roles in chemical reactions, biological processes, and material properties. In this exploration, we delve deeper into the characteristics, implications, and significance of these essential types of chemical bonding.

First, let's define polar and nonpolar molecules: A polar molecule occurs when there is an unequal sharing of electrons within a bond. A nonpolar molecule occurs when there is equal sharing of electrons within a bond.

3.5.2 Quick Ways to Distinguish Polar and Non-Polar Molecules

Let's first see some ways to distinguish nonpolar and polar molecules. Whenever you see a monoatomic atom such as helium, neon, argon, and so on, it is nonpolar. Whenever you see diatomic molecule of two identical atoms, such as nitrogen molecule (N2), chlorine molecule (Cl2), hydrogen molecule(H2), and so on, there is equal sharing of electrons between the atoms involved, therefore, they are a nonpolar. Another way to distinguish nonpolar molecules from polar molecules is to see if the compound is only composed of carbon and hydrogen atoms. So, whenever you see hydrocarbons, such as methane, ethane, benzene, and so on, it is a nonpolar molecule. But when you're dealing with compounds that are not diatomic, monoatomic, or even hydrocarbons, check to see if the compound is has a symmetric distribution of charges. Such a compound typically has a central atom that bonds all of its valence electrons equally to identical terminal atoms. Let's take carbon dioxide as an example. Since the central atom, carbon, bonds all 4 of its valence electrons equally to two terminal atoms, oxygen, it has a symmetric charge distribution, and is therefore nonpolar.

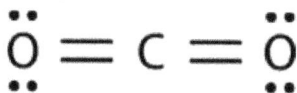

$$\overset{\bullet\bullet}{\underset{\bullet\bullet}{O}} = C = \overset{\bullet\bullet}{\underset{\bullet\bullet}{O}}$$

Figure 1: Carbon dioxide Lewis

Structure Another example is carbon tetrafluoride in which carbon as the central atom uses all of its valence electrons to share equal bonds with 4 fluorine atoms. This compound has a symmetrical distribution of charges, and so it's nonpolar.

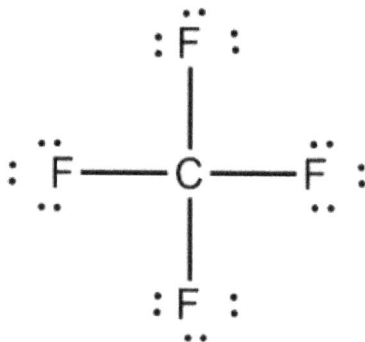

Figure 2: Carbon tetrafluoride Lewis Structure

One last example is boron trifluoride. The central atom boron uses all 3 of its valence electrons to share equal bonds with 3 fluorine atoms. So, this compound is also non-polar.

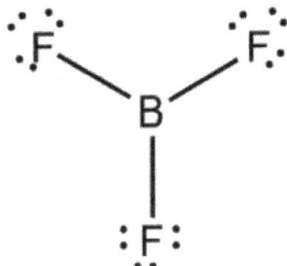

Figure 3: Boron trifluoride Lewis Structure Calculating the electronegativity difference between the bonded molecules can also help us determine whether a compound is nonpolar or polar. If the electronegativity difference is less than 0.4, then the substance is nonpolar. For example, in the compound iodine monobromide, Iodine has an electronegativity value of 2.66 while bromine has an electronegativity value of 2.96. The electronegativity difference between these two atoms is 0.3 which is less than 0.4, therefore, iodine monobromide is a nonpolar molecule. We can also identify nonpolar molecules by checking to see if there is zero net dipole moment in the compound. Take trans-1,2-dibromoethene for example. In this compound, two carbons are bonded to bromine and hydrogen atoms in trans-configuration.

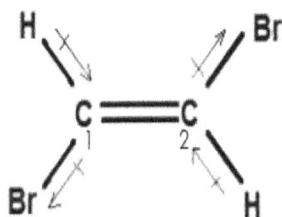

Figure 4: trans-1,2-dibromoethene Structural Formula Bromine is a more electronegative element compared to carbon, therefore the bond electrons shared between them are more attracted to bromine making the dipole moment in the direction towards bromine. While the bromine bonded to first carbon is attracting the bonded electrons towards itself, the bromine on second carbon is doing the same making the dipole moment is in the opposite direction. Therefore, dipole moment in these two bonds cancel each other out. In the bond between carbon and hydrogen, since carbon is more electronegative than hydrogen, it is going to attract the bonded electrons towards itself. Since the carbon-hydrogen bonds are in trans-configuration, the dipole moment in the bonds will cancel each other out. Therefore, overall, there is going to be zero net dipole moment making the entire compound nonpolar. Now let's see how to distinguish polar molecules from nonpolar molecules. First and foremost, if a compound has hydrogen bonding within it, it is a polar molecule. This means that if there is a hydrogen atom directly bonded to fluorine, nitrogen, or oxygen, then it is a polar

molecule. Even when it comes to methanol (CH3OH) where the methyl (CH3-) part does not have H-bonding, the entire compound is still considered polar because the hydroxyl group (-OH) involves H-bonding. If the compound you're looking at has an asymmetric charge distribution, then it is a polar molecule. The compound methyl chloride, CH3Cl, has three hydrogen atoms and one chlorine atom bonded to the central carbon. Since carbon is not bonded to identical terminal atoms, there is an asymmetric charge distribution in which most of the electrons will concentrate around chlorine as it is the most electronegative element in the compound. This makes it a polar compound.

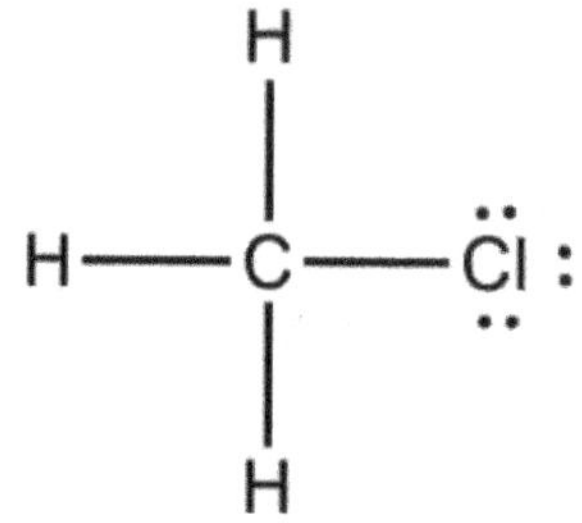

Figure 5: Chloromethane Lewis Structure

Similarly, Trichlorofluorosilane (SiFCl3) has an asymmetric charge distribution because one fluorine atom and three chlorine atoms are bonded to the central atom silicon. There will not be equal sharing of electrons because the terminal atoms are not identical. Thus, this compound is polar.

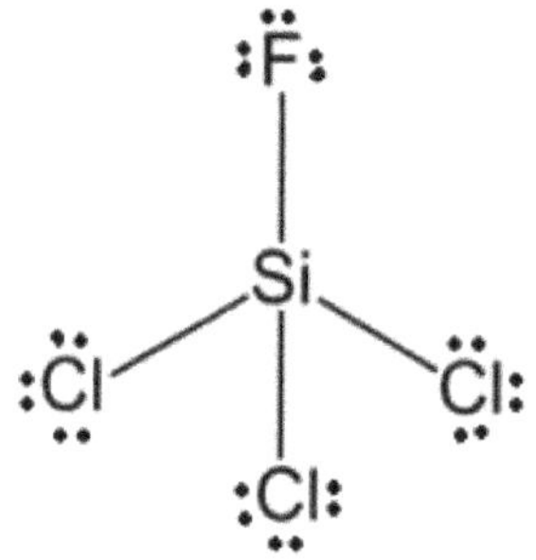

Figure 6: Trichlorofluorosilane Lewis Structure

Let's take a look at sulfur dioxide (SO2). You need to draw out the Lewis structure to see if its charge distribution is symmetrical or not. Sulfur forms a double bond with each oxygen and will be left with a lone pair. This molecule will have a bent molecular shape because of the lone pair on the central atom sulfur. The charge distribution is not symmetrical, making it a polar molecule.

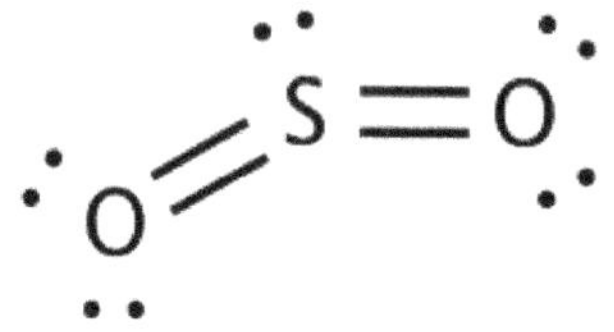

Figure 7: Sulfur dioxide Lewis Structure

If you see two atoms with great electronegativity difference bonded together, it is most likely a polar molecule. Whenever the electronegativity difference between two covalently bonded atoms is greater than 0.4, it is a polar molecule. Let's look at sulfur dioxide (SO_2) as an example, Sulfur's electronegativity is 2.58 and oxygen's electronegativity is 3.44. The electronegative difference between S and O is 0.86, so this molecule is surely polar. Go ahead and find the electronegativity difference between selenium and chlorine. If the value you find is greater than 0.4 but less than 1.8, then it should be a polar molecule. If the electronegativity difference happens to be greater than 1.8, then the compound you're looking at is most likely formed from an ionic bond. But if the difference is less than 0.4, then it is probably a non-polar molecule.

A final way I'm going to show you how to identify a polar molecule is by determining the net dipole moment of a compound. If it is not zero, then it is a polar molecule. Let's look at cis-dibromoethene as an example.

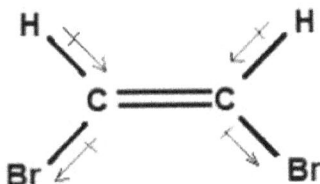

Figure 8: cis-dibromoethene Structural Formula

In the bond between carbon and bromine, since bromine is the more electronegative element, the dipole moment of the bond is in the direction towards bromine. Both carbon-bromine bonds are on the same side of the compound, and therefore the dipole moments in the bonds are not in opposite direction, and so they won't cancel each other out. Similarly, since the carbonhydrogen bonds are on the same side of the compound, then the dipole moments are not in opposite directions, so they cannot cancel each other out. Therefore, this compound has a non-zero dipole moment, thus, it is polar.

3.6 MULTIPLE BONDING: DOUBLE AND TRIPLE BONDS

3.6.1 Introduction

Multiple bonding, characterized by double and triple bonds, stands as a cornerstone in the language of chemical structures, enriching the diversity and reactivity of molecules across various disciplines of chemistry. These bonds, which involve the sharing of electron pairs between atoms, extend beyond the realm of single bonds to create enhanced connectivity and distinct chemical behaviors. Double bonds, featuring two shared electron pairs, and triple bonds, with three such pairs, emerge as pivotal components in understanding molecular geometry, reactivity patterns, and the functional roles of compounds in organic, inorganic, and biochemical contexts. From the fundamental principles of covalent bonding to the intricate mechanisms of organic synthesis, the exploration of multiple bonding unveils a fascinating interplay of electronic structure, molecular energetics, and chemical reactivity, offering profound insights into the intricate tapestry of molecular architecture and behavior. In this discourse, we delve into the structural characteristics, bonding nature, and transformative implications of double and triple bonds, elucidating their significance across the vast landscape of chemical inquiry and innovation.

DOUBLE BONDS

The Lewis structure of ethene, C_2H_4, shows us that each carbon atom is surrounded by one other carbon atom and two hydrogen atoms.

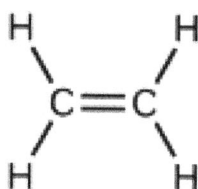

Figure.

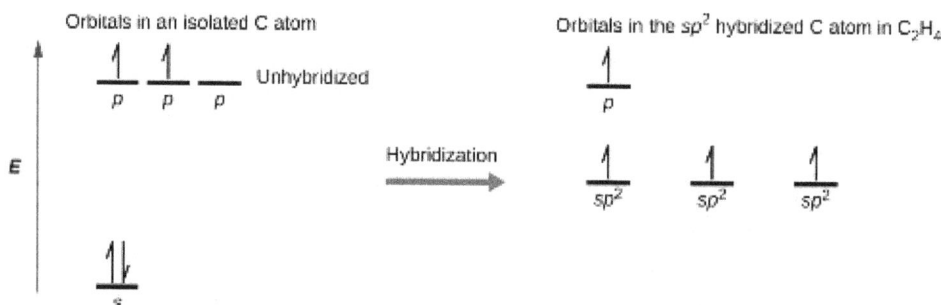

Figure : In ethene, each carbon atom is sp2 hybridized, and the sp2 orbitals and the p orbital are singly occupied. The hybrid orbitals overlap to form σ bonds, while the p orbitals on each carbon atom overlap to form a π bond.

The π bond in the C=C double bond results from the overlap of the third (remaining) $2p$ orbital on each carbon atom that is not involved in hybridization. This unhybridized p orbital is perpendicular to the plane of the $sp2$ hybrid orbitals. Thus the unhybridized $2p$ orbitals overlap in a side-by-side fashion, above and below the internuclear axis and form a π bond.

In an ethene molecule, the four hydrogen atoms and the two carbon atoms are all in the same plane. If the two planes of $sp2$ hybrid orbitals tilted relative to each other, the p orbitals would not be oriented to overlap efficiently to create the π bond. The planar configuration for the ethene molecule occurs because it is the most stable bonding arrangement. This is a significant difference between σ and π bonds; rotation around single (σ) bonds occurs easily because the end-to-end orbital overlap does not depend on the relative orientation of the orbitals on each atom in the bond. In other words, rotation around the internuclear axis does not change the extent to which the σ bonding orbitals overlap because the bonding electron density is symmetric about the axis. Rotation about the internuclear axis is much more difficult for multiple bonds; however, this would drastically alter the off-axis overlap of the π bonding orbitals, essentially breaking the π bond.

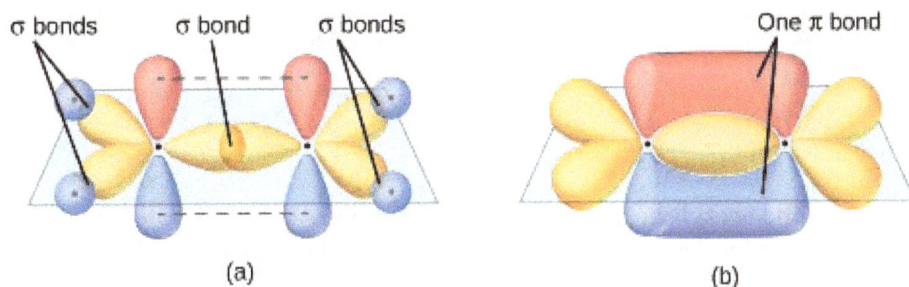

(a)   (b)

Figure : In the ethene molecule, C_2H_4, there are (a) five σ bonds. One C–C σ bond results from overlap of sp2 hybrid orbitals on the carbon atom with one sp2 hybrid orbital on the other carbon atom. Four C–H bonds result from the overlap between the C atoms› sp2 orbitals with s orbitals on the hydrogen atoms. (b) The π bond is formed by the side-by-side overlap of the two unhybridized p orbitals in the two carbon atoms. The two lobes of the π bond are above and below the plane of the σ system.

Triple Bonds

In molecules with sp hybrid orbitals, two unhybridized p orbitals remain on the atom (Figure).

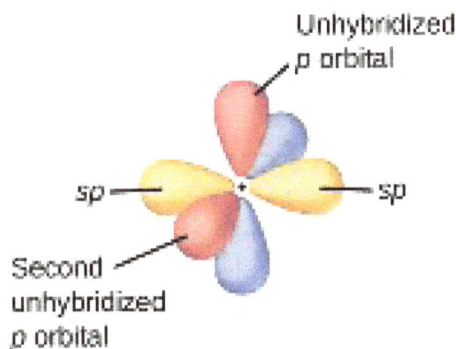

Figure : Diagram of the two linear sp hybrid orbitals of a carbon atom, which lie in a straight line, and the two unhybridized p orbitals at perpendicular angles.

We find this situation in acetylene, H–C≡C–H, which is a linear molecule. The sp hybrid orbitals of the two carbon atoms overlap end to end to form a σ bond between the carbon atoms (Figure). The remaining *sp* orbitals form σ bonds with hydrogen atoms. The two unhybridized p orbitals per carbon are positioned such that they overlap side by side and, hence, form two π bonds. The two carbon atoms of acetylene are thus bound together by one σ bond and two π bonds, giving a triple bond.

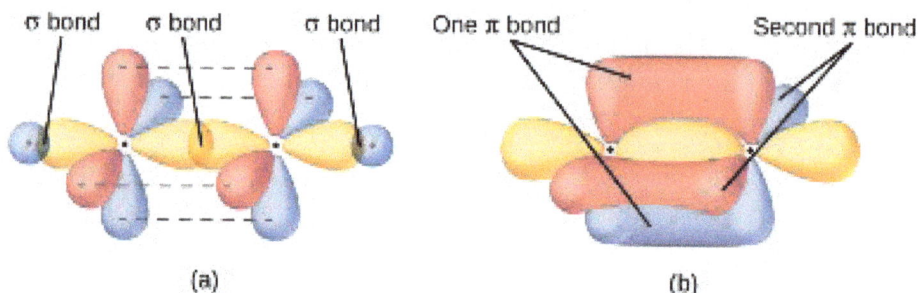

(a)   (b)

Figure : (a) In the acetylene molecule, C2H2, there are two C–H σ bonds and a C☐C triple bond involving one C–C σ bond and two C–C π bonds. The dashed lines, each connecting two lobes, indicate the side-by-side overlap of the four unhybridized p orbitals. (b) This shows the overall outline of the bonds in C2H2. The two lobes of each of the π bonds are positioned across from each other around the line of the C–C σ bond.

Application: sp3, sp2, or sp hybridization and bond type Double bonds and triple bonds require p-orbitals to make bonds. An sp3 hybrid orbital has used all of its p-orbitals, there are none left to be part of a multiple bond. An sp2 hybrid orbital has used two of three its p-orbitals, there is one left to work with a bond to create a double bond. An sp hybrid orbital has used one of three its p-orbitals, there are two left to work with a bond to create a triple bond. In general:

Only single bonds bond indicates sp3 hybridization. A double bond indicates sp2 hybridization.

A triple bond indicates sp hybridization.

Hybridization involves only σ bonds, lone pairs of electrons, and single unpaired electrons (radicals). Structures that account for these features describe the correct hybridization of the atoms. However, many structures also include resonance forms. Remember that resonance forms occur when various arrangements of π bonds are possible. Since the arrangement of π bonds involves only the unhybridized orbitals, resonance does not influence the assignment of hybridization.

For example, molecule benzene has two resonance forms (Figure). We can use either of these forms to determine that each of the carbon atoms is bonded to three other atoms with no lone pairs, so the correct hybridization is *sp2*. The electrons in the unhybridized p orbitals form π bonds. Neither resonance structure completely describes the electrons in the π bonds. They are not located in one position or the other, but in reality are delocalized throughout the ring. Valence bond theory does not easily address delocalization. Bonding in molecules with resonance forms is better described by molecular orbital theory. of Hybridization Involving Resonance

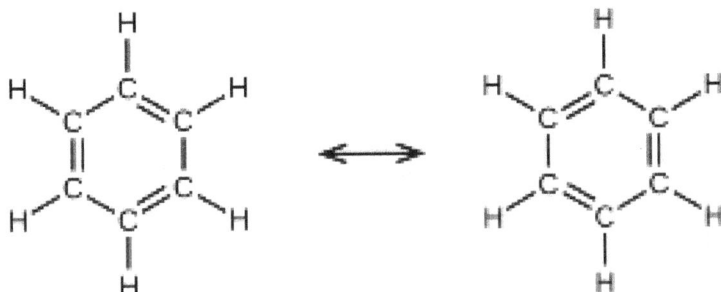

Figure : Each carbon atom in benzene, C6H6, is sp2 hybridized, independently of which resonance form is considered. The electrons in the π bonds are not located in one set of p orbitals or the other, but rather delocalized throughout the molecule.

Some acid rain results from the reaction of sulfur dioxide with atmospheric water vapor, followed by the formation of sulfuric acid. Sulfur dioxide, , is a major component of volcanic gases as well as a product of the combustion of sulfur-containing coal. What is the hybridization of the atom in ?

Solution

The resonance structures of are:

The sulfur atom is surrounded by two bonds and one lone pair of electrons in either resonance structure. Therefore, the electron-pair geometry is trigonal planar, and the hybridization of the sulfur atom is *sp2*.

Exercise

Another acid in acid rain is nitric acid, HNO3, which is produced by the reaction of nitrogen dioxide, NO2, with atmospheric water vapor. What is the hybridization of the nitrogen atom in NO2?

(**Note:** the lone electron on nitrogen occupies a hybridized orbital just as a lone pair would.)

sp2

More Than One Multiple Bond On The Central Atom

The central atom in a molecule can have more than one multiple bond. For example, each carbon atom in carbon dioxide has two unhybridized atomic *p* orbitals, and each oxygen atom still has one p orbital available. When the two O-atoms are brought up to opposite sides of the carbon atom, one of the p orbitals on each oxygen forms a π bond with one of the carbon p-orbitals. In this case, sp-hybridization is seen to lead to two double bonds (Figure). Notice that the two C–O π bonds are mutually perpendicular.

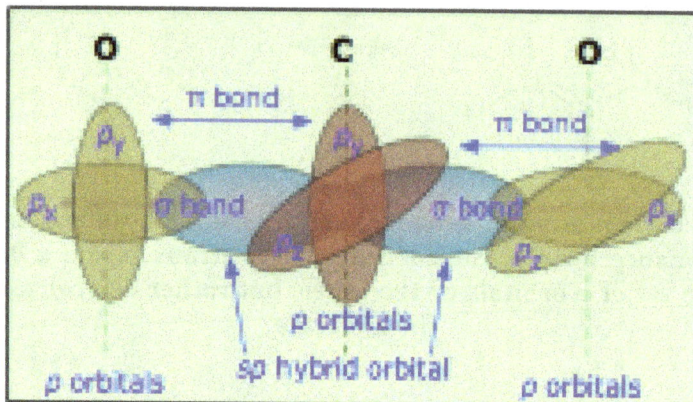

Figure : The sp-hybrized carbon forms two double bonds in carbon dioxide.

Summary

Multiple bonds consist of a σ bond located along the axis between two atoms and one or two π bonds. The σ bonds are usually formed by the overlap of hybridized atomic orbitals, while the π bonds are formed by the side-by-side overlap of unhybridized orbitals. Resonance occurs when there are multiple unhybridized orbitals with the appropriate alignment to overlap, so the placement of π bonds can vary.

04 | METALLIC BONDING

4.1 INTRODUCTION

Metallic bonding stands as one of the foundational pillars of materials science, providing the cornerstone for understanding the properties and behaviors of a vast array of metals and alloys. At its essence, metallic bonding embodies the captivating interplay between electrons and atomic nuclei within metallic structures, shaping the unique characteristics that define metals as a distinct class of materials.Central to the concept of metallic bonding is the notion of electron delocalization. Unlike the localized electron arrangements found in covalent or ionic bonds, metallic bonding involves the collective sharing of valence electrons among a multitude of atoms within a metal lattice. This shared electron sea forms a cohesive force that binds the positively charged metal ions together, giving rise to the remarkable properties exhibited by metallic substances. The nature of metallic bonding endows metals with an extraordinary combination of traits, including high electrical and thermal conductivity, malleability, ductility, and luster. These properties stem from the free movement of electrons throughout the material, facilitating the efficient transfer of energy and the deformation of the metal lattice without shattering. Moreover, metallic bonding engenders a degree of plasticity that enables metals to be shaped, formed, and molded into a myriad of shapes and configurations, making them indispensable in various industrial applications. From the towering skyscrapers that define modern cityscapes to the intricate machinery that drives technological innovation, metallic bonding underpins the structural integrity and functional versatility of countless engineering marvels.

Beyond its practical implications, the study of metallic bonding serves as a gateway to exploring the fundamental principles governing the behaviour of matter at the atomic and molecular levels. Through theoretical models, computational simulations, and experimental observations, scientists delve into the intricacies of electron dynamics, crystal structures, and phase transitions, deepening our understanding of the complex phenomena that underlie the macroscopic properties of metals. In essence, metallic bonding encapsulates the elegant interplay of quantum mechanics and classical physics, offering insights into the nature of matter and the boundless possibilities for harnessing its potential.

As researchers continue to unravel the mysteries of metallic bonding, they pave the way for transformative breakthroughs in materials science, engineering, and technology, shaping the trajectory of human progress for generations to come.

4.2 METALLIC BONDING CONCEPT

Metallic bonding is a fundamental concept in chemistry that underlies the unique properties of metals. Unlike other forms of bonding like ionic or covalent, metallic bonding occurs within metals and metalloids, defining their characteristic traits. At its core, metallic bonding involves the sharing of valence electrons among a lattice of metal atoms. These atoms release their outer electrons, creating a "sea" of delocalized electrons that move freely throughout the material. Meanwhile, the metal ions left behind form a closely packed lattice structure. This arrangement results in a strong electrostatic attraction between the positive metal ions and the negatively charged electron sea, holding the structure together. This bonding mechanism explains key properties of metals such as electrical conductivity, malleability, ductility, and luster, making them essential materials in various industries and everyday applications.

4.2.1 Metallic bonding in sodium

Metals tend to have high melting points and boiling points suggesting strong bonds between the atoms. Even a metal like sodium (melting point 97.8°C) melts at a considerably higher temperature than the element (neon) which precedes it in the Periodic Table.

Sodium has the electronic structure $1s^2 2s^2 2p^6 3s^1$. When sodium atoms come together, the electron in the 3s atomic orbital of one sodium atom shares space with the corresponding electron on a neighbouring atom to form a molecular orbital - in much the same sort of way that a covalent bond is formed.

The difference, however, is that each sodium atom is being touched by eight other sodium atoms - and the sharing occurs between the central atom and the 3s orbitals on all of the eight other atoms. And each of these eight is in turn being touched by eight sodium atoms, which in turn are touched by eight atoms - and so on and so on, until you have taken in all the atoms in that lump of sodium.

All of the 3s orbitals on all of the atoms overlap to give a vast number of molecular orbitals which extend over the whole piece of metal. There have to be huge numbers of molecular orbitals, of course, because any orbital can only hold two electrons.

The electrons can move freely within these molecular orbitals, and so each electron becomes detached from its parent atom. The electrons are said to be delocalised. The metal is held together by the strong forces of attraction between the positive nuclei and the delocalised electrons.

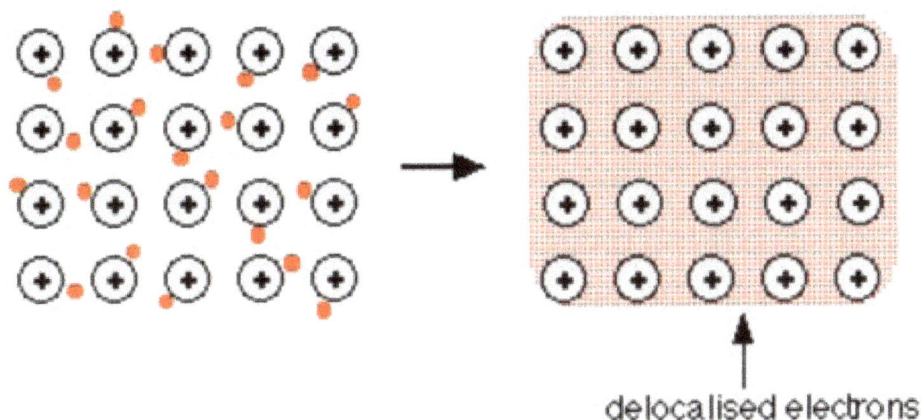

delocalised electrons

This is sometimes described as "an array of positive ions in a sea of electrons".

If you are going to use this view, beware! Is a metal made up of atoms or ions? It is made of atoms.

Each positive centre in the diagram represents all the rest of the atom apart from the outer electron, but that electron hasn't been lost - it may no longer have an attachment to a particular atom, but it's still there in the structure. Sodium metal is therefore written as Na⁻ not Na⁺.

4.2.2 Metallic bonding in magnesium

If you work through the same argument with magnesium, you end up with stronger bonds and so a higher melting point.

Magnesium has the outer electronic structure 3s2. Both of these electrons become delocalised, so the "sea" has twice the electron density as it does in sodium. The remaining "ions" also have twice the charge (if you are going to use this particular view of the metal bond) and so there will be more attraction between "ions" and "sea".

More realistically, each magnesium atom has one more proton in the nucleus than a sodium atom has, and so not only will there be a greater number of delocalised electrons, but there will also be a greater attraction for them.

Magnesium atoms have a slightly smaller radius than sodium atoms, and so the delocalised electrons are closer to the nuclei. Each magnesium atom also has twelve near neighbours rather than sodium's eight. Both of these factors increase the strength of the bond still further.

4.2.3 Metallic bonding in transition elements

Transition metals tend to have particularly high melting points and boiling points. The reason is that they can involve the 3d electrons in the delocalisation as well as the 4s. The more electrons you can involve, the stronger the attractions tend to be.

4.2.4 The metallic bond in molten metals

In a molten metal, the metallic bond is still present, although the ordered structure has been broken down. The metallic bond isn't fully broken until the metal boils. That means that boiling point is actually a better guide to the strength of the metallic bond than melting point is. On melting, the bond is loosened, not broken.

4.3 METALLIC STRUCTURES

4.3.1 The arrangement of the atoms

Metals are giant structures of atoms held together by metallic bonds. "Giant" implies that large but variable numbers of atoms are involved - depending on the size of the bit of metal.

4.3.2 12-co-ordination

Most metals are close packed - that is, they fit as many atoms as possible into the available volume. Each atom in the structure has 12 touching neighbours. Such a metal is described as 12-co-ordinated.

Each atom has 6 other atoms touching it in each layer.

There are also 3 atoms touching any particular atom in the layer above and another 3 in the layer underneath.

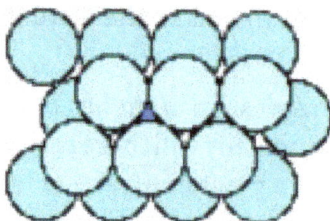

This second diagram shows the layer immediately above the first layer. There will be a corresponding layer underneath. (There are actually two different ways of placing the third layer in a close packed structure, but that goes beyond the requirements of current A'level syllabuses.)

4.3.3 8-co-ordination

Some metals (notably those in Group 1 of the Periodic Table) are packed less efficiently, having only 8 touching neighbors. These are 8-co-ordinated.

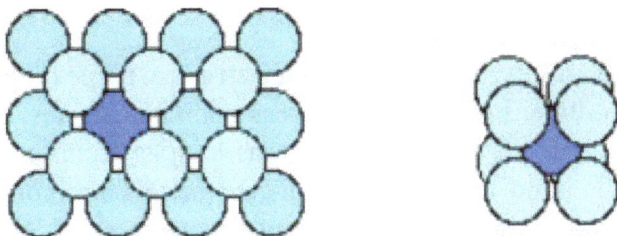

The left hand diagram shows that no atoms are touching each other within a particular layer . They are only touched by the atoms in the layers above and below. The right hand diagram shows the 8 atoms (4 above and 4 below) touching the darker coloured one.

4.3.4 Crystal grains

It would be misleading to suppose that all the atoms in a piece of metal are arranged in a regular way. Any piece of metal is made up of a large number of "crystal grains", which are regions of regularity. At the grain boundaries atoms have become misaligned.

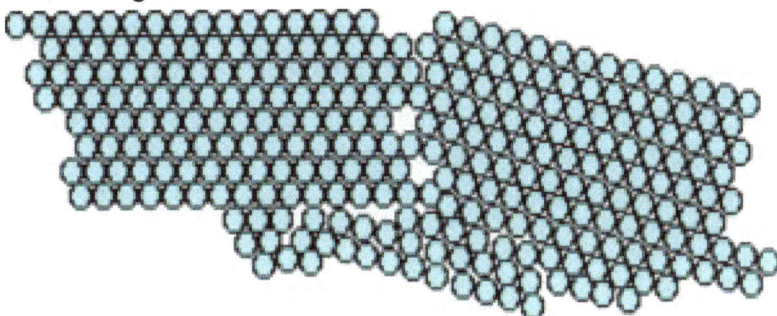

4.4 PHYSICAL PROPERTIES OF METALS

4.4.1 Melting points and boiling points

Metals tend to have high melting and boiling points because of the strength of the metallic bond. The strength of the bond varies from metal to metal and depends on the number of electrons which each atom delocalises into the sea of electrons, and on the packing.

Group 1 metals like sodium and potassium have relatively low melting and boiling points mainly because each atom only has one electron to contribute to the bond - but there are other problems as well:

⊙ Group 1 elements are also inefficiently packed (8-co-ordinated), so that they aren't forming as many bonds as most metals.

⊙ They have relatively large atoms (meaning that the nuclei are some distance from the delocalised electrons) which also weakens the bond.

4.4.2 Electrical conductivity

Metals conduct electricity. The delocalised electrons are free to move throughout the structure in 3-dimensions. They can cross grain boundaries. Even though the pattern may be disrupted at the boundary, as long as atoms are touching each other, the metallic bond is still present. Liquid metals also conduct electricity, showing that although the metal atoms may be free to move, the delocalisation remains in force until the metal boils.

4.4.3 Thermal conductivity

Metals are good conductors of heat. Heat energy is picked up by the electrons as additional kinetic energy (it makes them move faster). The energy is transferred throughout the rest of the metal by the moving electrons.

4.4.4 Strength and workability

Malleability and ductility

Metals are described as malleable (can be beaten into sheets) and ductile (can be pulled out into wires). This is because of the ability of the atoms to roll over each other into new positions without breaking the metallic bond.

If a small stress is put onto the metal, the layers of atoms will start to roll over each other. If the stress is released again, they will fall back to their original positions. Under these circumstances, the metal is said to be elastic.

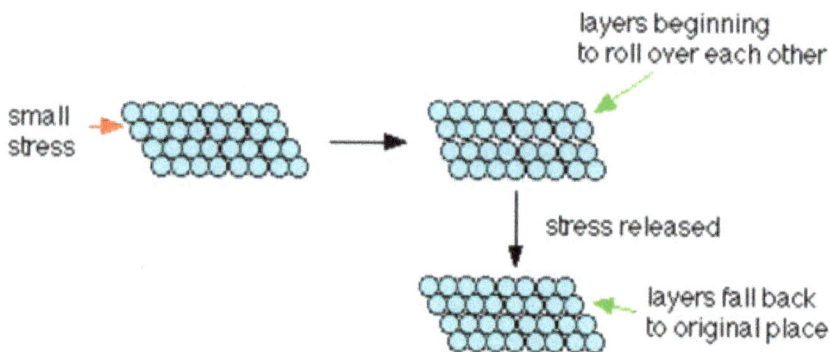

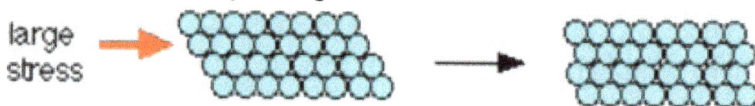

The hardness of metals

This rolling of layers of atoms over each other is hindered by grain boundaries because the rows of atoms don't line up properly. It follows that the more grain boundaries there are (the smaller the individual crystal grains), the harder the metal becomes.

Offsetting this, because the grain boundaries are areas where the atoms aren't in such good contact with each other, metals tend to fracture at grain boundaries. Increasing the number of grain boundaries not only makes the metal harder, but also makes it more brittle.

Controlling the size of the crystal grains

If you have a pure piece of metal, you can control the size of the grains by heat treatment or by working the metal.

Heating a metal tends to shake the atoms into a more regular arrangement - decreasing the number of grain boundaries, and so making the metal softer. Banging the metal around when it is cold tends to produce lots of small grains. Cold working therefore makes a metal harder. To restore its workability, you would need to reheat it.

You can also break up the regular arrangement of the atoms by inserting atoms of a slightly different size into the structure. Alloys such as brass (a mixture of copper and zinc) are harder than the original metals because the irregularity in the structure helps to stop rows of atoms from slipping over each other.

4.5 ELECTRON SEA MODEL

German physicist, Paul Drude, postulated a theory, or say framework called Electron Sea Model. This theory is most often used in understanding bonding in metals. The 'Electron Sea model' framework is also known as 'Electron gas theory.' Moreover, the theory focuses on revealing metals' unique properties, such as the free nature of valence ions of metals when they do bonding. Furthermore, Henrik Lorentz also contributed to the development of this theory.

Electron Sea Model Meaning – A model depicts the free nature of electrons around the metal atom. According to this concept, any of these metal atoms in a metal solid donate its valence electron to create an electron "sea." These electrons inside the connecting metallic atoms' outer energy levels are not bound by anything and can readily migrate from one atom to another. The purpose of the Electron Sea Model is to explain the electrical and thermal conductivity of metals by screening the nature of their electrons.

Metallic Bonding

The entire theory of the Electron Sea Model revolves around the nature of atoms during this bonding. Metallic bonding is the exchange of unpaired electrons across positively energised metal ions in a mesh. Metal bonds differ significantly from covalent or ionic bonds in their structure. Metallic bonds are accountable for the connection among metal atoms, whereas ionic bonds connect metals to nonmetals and though instead, covalent bonds connect non-metals to nonmetals.

4.5.1 Relationship between Electron Sea Model and Properties of Metallic bonding

The Electron Sea Model shows many physical and chemical properties of metallic bonding. Some of them are as follows.

- ◉ Whenever an electric field, such as from a battery, is introduced, these moving electrons can conduct electrical change. Regardless of the drifting electrons and the ease with which the cations roll through each other, they are flexible.

- ◉ Since metals have free electrons, they reflect light generally. Furthermore, as per the model, it has a double number of electrons as the sole electrons as two 3s electrons are delocalised and get free. Because electrons move readily in them, they serve as good electrical conductors. It indicates that in metallic interaction, the atom would shed the electron concentration in metallic interaction despite the electron having delivered to another atom to remain more stable.

4.5.2 Statements of Electron Sea Model

- ◉ The valence electrons of metals aren't even bound tightly by the nuclei due to the lower ionisation energies, and the electrons can travel freely utilising unoccupied orbitals.

- ◉ Metal atoms rapidly shed their valence electrons. The kernel is a collection of positive charge ions which they can generate.

- ◉ Positively charged ions are grouped in a geometric pattern. This cluster of metal ions, or kernel, is immersed in a sea of simply moving electrons.

- ◉ The mobile electrons freely travel around the collection of metal ions, much like seawater or any other gas. As a result, it's also known as the electron gas theory.

Conclusion

We conclude that the electron sea model is a system by which the electrons of atoms of different elements work. The theory was proposed in the early 1900s. Moreover, the model states atoms tend to have free moving molecules around the atomic nuclei. As a result, many of the properties of elements can be understood by the model.

4.6 ALLOYS AND THEIR SIGNIFICANCE

4.6.1 Alloys

Alloy plays a very important role in our daily life. Utensils in the kitchen, vehicles, mobile phones etc are various alloys which are being used and made by human. Here, we are providing the list of important alloys and their uses which will help in revising at the time of examination.

Alloy plays a very important role in our daily life. In fact without using an alloy one day is also not passed. Utensils in the kitchen, vehicles, mobile phones etc are various alloys which are being used and made by human. Even most of the machinery tools and engineering equipments are made up of alloys.

4.6.2 Properties of an Alloy

Metal Alloys have both physical and chemical properties together with mechanical. Some properties are reactivity, electrical conductivity, thermal conductivity, good tensile strength, resistance to deformation, malleability etc.

4.6.3 Significance of Alloys

WAlloys are of immense significance across various fields due to their unique properties and versatility. Here are some key points highlighting the significance of alloys:

- ◉ *Strength and Durability:* Alloys often possess superior strength and durability compared to pure metals. They can withstand higher stresses and strains, making them suitable for structural applications in construction, automotive, aerospace, and other industries.

- ◉ *Corrosion Resistance:* Many alloys exhibit excellent resistance to corrosion, oxidation, and other forms of environmental degradation. This property makes them ideal for use in harsh or corrosive environments, such as marine applications and chemical processing plants.

- ◉ *Enhanced Mechanical Properties:* Alloys can be engineered to have specific mechanical properties, including hardness, ductility, toughness, and fatigue resistance. This allows manufacturers to tailor materials to meet the demands of particular applications.

- ◉ *Electrical and Thermal Conductivity:* Some alloys, such as copper alloys, offer high electrical and thermal conductivity, making them invaluable in electrical wiring, electronics, heat exchangers, and other heat transfer applications.

- ◉ *Lightweight Construction:* Alloys can be formulated to be lightweight while maintaining strength and durability. This property is particularly crucial in industries like aerospace and transportation, where reducing weight can lead to increased fuel efficiency and improved performance.

- ◉ **Temperature Resistance:** Certain alloys, known as superalloys, are engineered to withstand extreme temperatures and harsh conditions. These alloys are used in gas turbines, jet engines, nuclear reactors, and other high-temperature applications.

- ◉ **Customization and Versatility:** The composition of alloys can be adjusted to achieve specific properties, allowing for a wide range of applications across diverse industries. This versatility enables engineers and manufacturers to meet various performance requirements and design specifications.

- ◉ **Cost-Effectiveness:** Alloys often offer a cost-effective solution compared to using pure metals or alternative materials. By blending different elements, manufacturers can create materials with desired properties at a reasonable cost.

- ◉ **Recyclability:** Alloys are generally recyclable, contributing to sustainable practices and reducing the demand for primary metal sources. Recycling alloys helps conserve resources and minimize environmental impact.

- ◉ **Innovation and Advancements:** Ongoing research and development in alloy design and processing techniques continue to drive innovation and advancements in materials science and engineering. New alloy formulations are continually being developed to address emerging challenges and opportunities in various industries.

In conclusion, alloys play a critical role in modern technology, infrastructure, and everyday life, offering a diverse range of properties and applications that contribute to advancements in numerous fields. Their significance lies in their ability to meet specific performance requirements, enhance efficiency, and drive progress in various industries.

List of Important Alloys and their Uses

Alloys	Compositions	Uses
Brass	Cu + Zn	In making utensils.
Bronze	Cu + Sn	In making coins, bell and utensils.
German Silver	Cu + Zn + Ni	In making utensils.
Rolled Gold	Cu + Al	In making cheap ornaments.
Gun Metal	Cu + Sn + Zn + Pb	In making guns, barrels, gears and bearings.
Dutch metal	Cu + Zn	In making artificial ornaments.
Delta metal	Cu + Zn + Fe	In making blades of aeroplane.

Alloys	Compositions	Uses
Munz metal	Cu + Zn	In making coins.
Monel metal	Cu + Ni	For base containing container.
Rose metal	Bi + Pb + Sn	For making automatic fuse.
Duralumin	Al + Cu + Mg + Mn	For making utensils.
Magnalium	Al + Mg	For frame of aeroplane.
Solder	Pb + Sn	For soldering.
Type metal	Sn + Pb + Sb	In printing industry.
Bell metal	Cu + Sn	For casting bells and statues.
Stainless steel	Fe + Cr + Ni + C	For making utensils and surgical cutlery.
Nickel steel	Fe + Ni	For making electrical wire, automobile parts.

5.1 ATOMIC ORBITAL HYBRIDIZATION

Atomic orbital hybridization is a fundamental concept in chemistry that plays a crucial role in understanding the structure, bonding, and properties of molecules. It involves the mixing of atomic orbitals to form new hybrid orbitals that better describe the geometry and bonding observed in molecules. In classical atomic theory, electrons were thought to occupy fixed orbits around the nucleus of an atom, each orbit corresponding to a specific energy level. These orbits were described using mathematical functions known as atomic orbitals, which provided a way to visualize the probability distribution of finding an electron within a certain region of space around the nucleus. However, when atoms combine to form molecules, the simple atomic orbitals of the individual atoms must reorganize to accommodate the bonding requirements and geometry of the molecule. This reorganization is achieved through hybridization, where atomic orbitals mix to form new hybrid orbitals. The most common types of hybridization encountered in organic chemistry are sp, sp2, and sp3 hybridization. Each type of hybridization results from the combination of different numbers and types of atomic orbitals.

5.1.1. sp Hybridization: In sp hybridization, one s orbital and one p orbital from the valence shell of an atom combine to form two new hybrid orbitals. These hybrid orbitals are called sp hybrid orbitals. The name "sp" signifies the mixture of one s orbital and one p orbital. The sp hybrid orbitals are oriented linearly along a straight line, with an angle of 180 degrees between them. This linear arrangement is ideal for the formation of linear molecules. Examples of molecules exhibiting sp hybridization include beryllium chloride ($BeCl2$) and carbon monoxide (CO).

5.1.2. sp^2 Hybridization: In sp^2 hybridization, one's orbital and two p orbitals from the valence shell of an atom combine to form three new hybrid orbitals called sp^2 hybrid orbitals. The name "sp^2" indicates the mixture of one's orbital and two p orbitals.

The sp^2 hybrid orbitals are arranged in a trigonal planar geometry, with bond angles of approximately 120 degrees between them. This geometry is characteristic of molecules with a flat, triangular shape. Examples of molecules exhibiting sp^2 hybridization include ethylene (C_2H_4) and boron trifluoride (BF_3).

5.1.3. sp³ Hybridization: In sp3 hybridization, one's orbital and three p orbitals from the valence shell of an atom combine to form four new hybrid orbitals called sp3 hybrid orbitals. The name "sp3" signifies the mixture of one's orbital and three p orbitals.

The sp³ hybrid orbitals adopt a tetrahedral geometry, with bond angles of approximately 109.5 degrees between them. This geometry is commonly observed in molecules with four electron groups around the central atom. Examples of molecules exhibiting sp3 hybridization include methane (CH_4) and ethane (C_2H_6).

In summary, atomic orbital hybridization provides a powerful framework for understanding the shapes, bonding, and reactivity of molecules in chemistry. By considering hybridization, chemists can predict molecular geometries, identify bond angles, and rationalize the observed properties of a wide range of chemical compounds.

5.2 SP, SP², SP³ HYBRIDIZATION

5.2.1 sp3 HYBRIDIZATION

Definition

In *sp3* hybridization, the *2s* orbital is mixed with all three of the *2p* orbitals to give a set of four *sp3* hybrid orbitals. (The number of hybrid orbitals must equal the number of original atomic orbitals used for mixing.) The hybrid orbitals will each have the same energy but will be different in energy from the original atomic orbitals. That energy difference will reflect the mixing of the respective atomic orbitals. The energy of each hybrid orbital is greater than the original *s* orbital but less than the original *p* orbitals (*Fig. 1*).

Electronic configuration

The valence electrons for carbon can now be fitted into the *sp3* hybridized orbitals (*Fig. 1*). There was a total of four electrons in the original 2s and 2p orbitals. The *s* orbital was filled and two of the *p* orbitals were half filled. After hybridization, there is a total of four hybridized *sp3* orbitals all of equal energy. By Hund's rule, they are all half filled with electrons which means that there are four unpaired electrons. Four bonds are now possible.

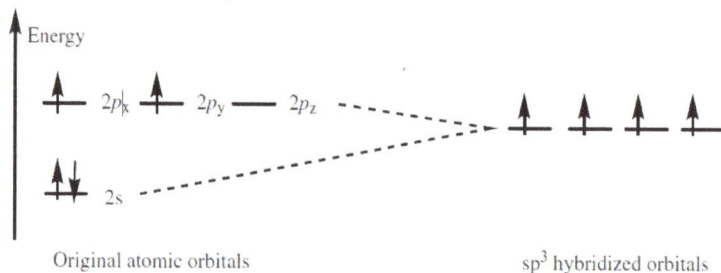

Fig. 1. sp3 Hybridization

Geometry

Each of the *sp3* hybridized orbitals has the same shape – a rather deformed looking dumbbell (*Fig. 2*). This deformed dumbbell looks more like a *p* orbital than an *s* orbital since more *p* orbitals were involved in the mixing process.

Minor lobe ⬭🟤 Major lobe

Fig. 2. sp3 Hybridized orbital.

Each *sp3* orbital will occupy a space as far apart from each other as possible by pointing to the corners of a tetrahedron (*Fig. 3*). Here, only the major lobe of each hybridized orbital has been shown and the angle between each of these lobes is 109.5°. This is what is meant by the expression **tetrahedral carbon**. The three-dimensional shape of the tetrahedral carbon can be represented by drawing a nor- mal line for bonds in the plane of the page. Bonds going behind the page are represented by a hatched wedge, and bonds coming out the page are represented by a solid wedge.

Tetrahedral shape

	Bond in the plane of the page
·····IIII	Bond going behind the page
◀	Bond coming out of the page

Fig. 3. Tetrahedral shape of an sp3 hybridized carbon

Sigma bonds

a)

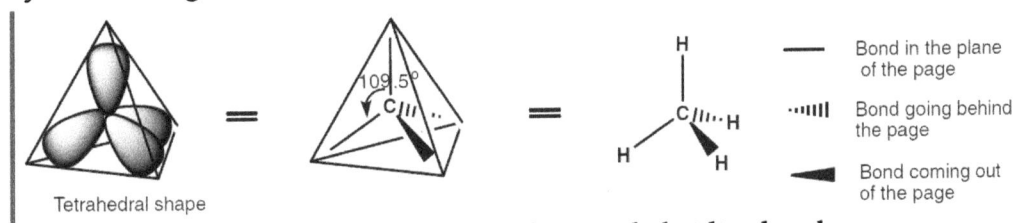

sp^3 sp^3 sigma bond

b)

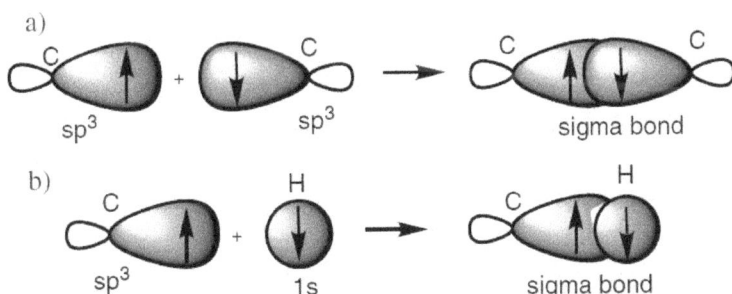

sp^3 1s   sigma bond

Fig. 4. (a) σ Bond between two sp3 hybridized carbons; (b) σ bond between an sp3hybridized carbon and hydrogen

A half-filled *sp3* hybridized orbital from one carbon atom can be used to form a bond with a half-filled *sp3* hybridized orbital from another carbon atom. In *Fig. 4a*, the major lobes of the two *sp3* orbitals overlap directly leading to a strong σ bond. It is the ability of hybridized orbitals to form strong σ bonds that explains why hybridization takes place in the first place. The deformed dumbbell shapes allow a much better orbital overlap than would be obtained from a pure *s* orbital or a pure *p* orbital. A σ bond between an *sp3* hybridized carbon atom and a hydrogen atom involves the carbon atom using one of its half-filled *sp3* orbitals and the hydrogen atom using its half-filled 1*s* orbital (*Fig. 4b*).

Nitrogen, oxygen, and chlorine

Nitrogen, oxygen and chlorine atoms can also be *sp3* hybridized in organic structures. Nitrogen has five valence electrons in its second shell. After hybridization, it will have three half-filled *sp3* orbitals and can form three bonds. Oxygen has six valence electrons. After hybridization, it will have two half-filled *sp3* orbitals and will form two bonds. Chlorine has seven valence electrons. After hybridization, it will have one half-filled *sp3* orbital and will form one bond.

The four *sp3* orbitals for these three atoms form a tetrahedral arrangement with one or more of the orbitals occupied by a lone pair of electrons. Considering the atoms alone, nitrogen forms a pyramidal shape where the bond angles are slightly less than 109.5° (c. 107°) (*Fig. 5a*). This compression of the bond angles is due to the orbital containing the lone pair of electrons, which demands a slightly greater amount of space than a bond. Oxygen forms an angled or bent shape where two lone pairs of electrons compress the bond angle from 109.5° to c. 104° (*Fig. 5b*).

Alcohols, amines, alkyl halides, and ethers all contain sigma bonds involving nitrogen, oxygen, or chlorine. Bonds between these atoms and carbon are formed by the overlap of half-filled *sp3* hybridized orbitals from each atom. Bonds involving hydrogen atoms (e.g. O–H and N–H) are formed by the overlap of the half-filled 1*s* orbital from hydrogen and a half-filled *sp3* orbital from oxygen or nitrogen.

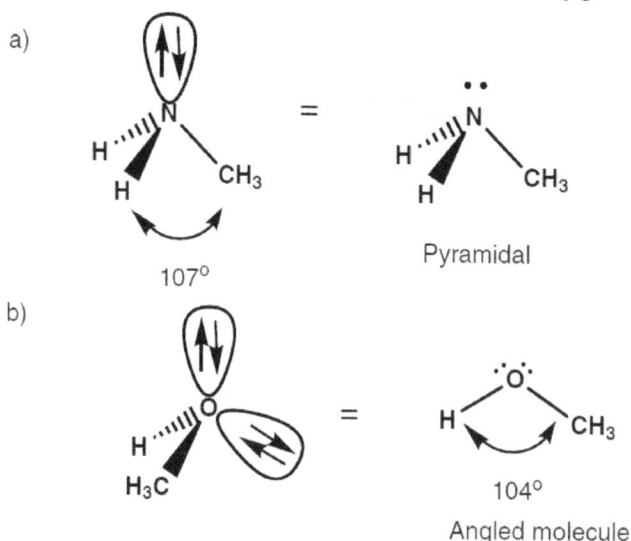

Fig. 5. (a) Geometry of sp3 hybridized nitrogen; (b) geometry of sp3 hybridized oxygen.

5.2.2 sp2 HYBRIDIZATION

Definition:

In *sp2* hybridization, the *s* orbital is mixed with two of the $2p$ orbitals (e.g. $2p_x$ and $2pz$) to give three *sp* hybridized orbitals of equal energy. The remaining $2py$ orbital is unaffected. The energy of each hybridized orbital is greater than the original s orbital but less than the original *p* orbitals. The remaining $2p$ orbital (in this case the $2py$ orbital) remains at its original energy level (*Fig. 1*).

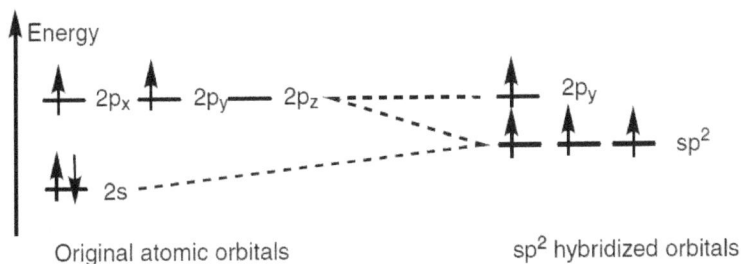

Fig. 1. sp2 Hybridization.

Electronic configuration

For carbon, there are four valence electrons to fit into the three hybridized *sp2* orbitals and the remaining $2p$ orbital. The first three electrons are fitted into each of the hybridized orbitals according to Hund's rule such that they are all half-y filled. This leaves one electron still to place. There is a choice between pairing it up in a half-filled *sp2* orbital or placing it into the vacant $2p$ orbital. The usual prin- ciple is to fill up orbitals of equal energy before moving to an orbital of higher energy. However, if the energy difference between orbitals is small (as here) it is easier for the electron to fit into the higher energy $2py$ orbital resulting in three half-filled *sp2* orbitals and one half-filled *p* orbital (*Fig. 1*). Four bonds are possible.

Geometry

The $2py$ orbital has the usual dumbbell shape. Each of the *sp* hybridized orbitals has a deformed dumbbell shape similar to an *sp3* hybridized orbital. However, the difference between the sizes of the major and minor lobes is larger for the *sp2* hybridized orbital.

The hybridized orbitals and the $2py$ orbital occupy spaces as far apart from each other as possible. The lobes of the $2py$ orbital occupy the space above and below the plane of the *x* and *z* axes (*Fig. 2a*). The three *sp2* orbitals (major lobes shown only) will then occupy the remaining space such that they are as far apart from the $2py$ orbital and from each other as possible. As a result, they are all placed in the *x–z* plane pointing toward the corner of a triangle (trigonal planar shape; *Fig. 2b*). The angle between each of these lobes is 120°. We are now ready to look at the bonding of an *sp2* hybridized carbon.

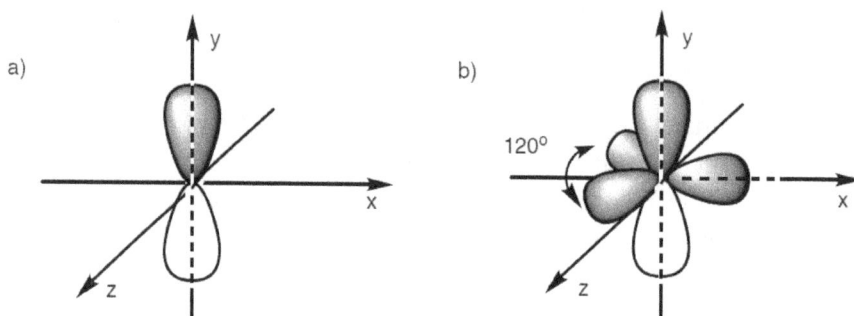

Fig. 2. (a) Geometry of the 2py orbital; (b) geometry of the 2 py orbital and the sp hybridized orbitals.

Alkenes

sp2 Hybridization results in three half-filled sp2 hybridized orbitals which form a trigonal planar shape. The use of these three orbitals in bonding explains the shape of an alkene, for example ethene (H2C=CH2). As far as the C–H bonds are concerned, the hydrogen atom uses a half-filled 1s orbital to form a strong σ bond with a half filled sp2 orbital from carbon (Fig. 3a). A strong σ bond is also possible between the two carbon atoms of ethene due to the overlap of sp2 hybridized orbitals from each carbon (Fig. 3b).

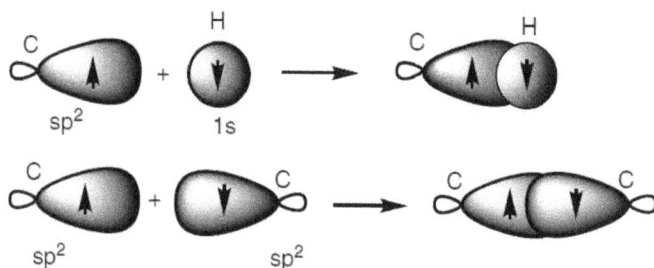

Fig. 3. (a) Formation of a C–H σ bond; (b) formation of a C–C σ bond.

The full σ bonding diagram for ethene is shown in Fig. 4a and can be simplified as shown in Fig. 4b. Ethene is a flat, rigid molecule where each carbon is trigonal planar. We have seen how sp2 hybridization explains the trigonal planar carbons but we have not explained why the molecule is rigid and planar. If the σ bonds were the only bonds present in ethene, the molecule would not remain planar since rotation could occur round the C–C σ bond (Fig. 5). Therefore, there must be further bonding which 'locks' the alkene into this planar shape. This bond involves

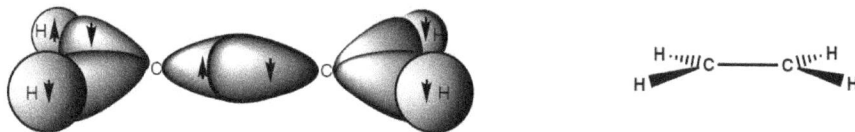

Fig. 4. (a) σ Bonding diagram for ethene; (b) simple representation of σ bonds for ethene.

Fig. 5. Bond rotation around a σ bond.

the remaining half-filled 2py orbitals on each carbon which overlap side-on to produce a **pi (p) bond**), with one lobe above and one lobe below the plane of the molecule (*Fig. 6*). This π bond prevents rotation round the C–C bond since the π bond would have to be broken to allow rotation. A π bond is weaker than a σ bond since the 2py orbitals overlap side-on, resulting in a weaker overlap. The presence of a π bond also explains why alkenes are more reactive than alkanes, since a π bond is more easily broken and is more likely to take part in reactions.

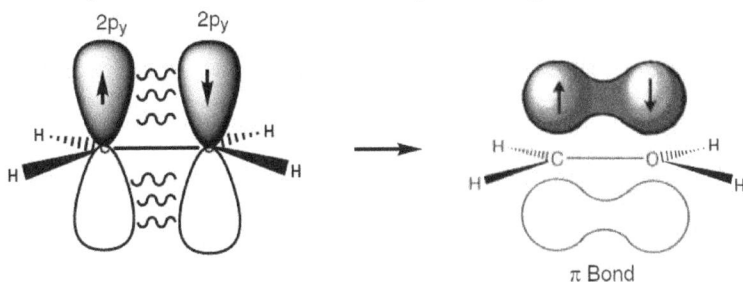

Fig. 6. Formation of a π bond.

Carbonyl groups:

The same theory explains the bonding within a carbonyl group (C=O) where both the carbon and oxygen atoms are $sp2$ hybridized. The following energy level diagram (*Fig. 7*) shows how the valence electrons of oxygen are arranged after $sp2$ hybridization. Two of the $sp2$ hybridized orbitals are filled with lone pairs of electrons, which leaves two half-filled orbitals available for bonding. The $sp2$ orbital can be used to form a strong σ bond, while the 2py orbital can be used for the weaker π bond. *Figure 8* shows how the σ and π bonds are formed in the carbonyl group and explains why carbonyl groups are planar with the carbon atom having a trigonal planar shape. It also explains the reactivity of carbonyl groups since the π bond is weaker than the σ bond and is more likely to be involved in reactions.

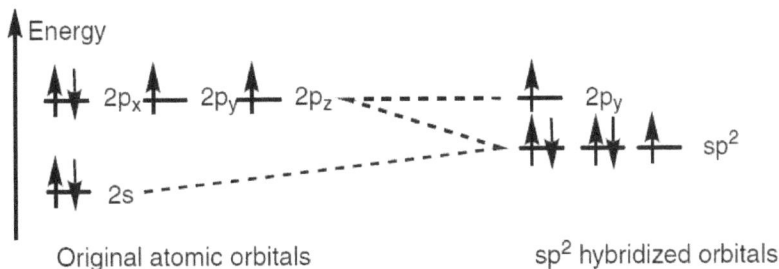

Fig. 7. Energy level diagram for sp2 hybridized oxygen.

a)

b)

2py   2py

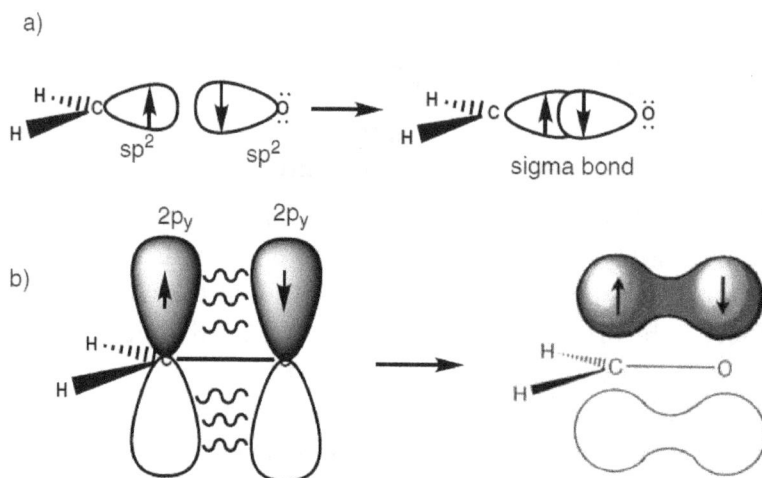

Fig. 8. (a) Formation of the carbonyl σ bond; (b) formation of the carbonyl π bond.

Aromatic rings:

All the carbons in an aromatic ring are *sp2* hybridized which means that each carbon can form three σ bonds and one π bond. In *Fig. 9a*, all the single bonds are σ while each double bond consists of one σ bond and one π bond. However, this is an oversimplification of the aromatic ring. For example, double bonds are shorter than single bonds and if benzene had this exact structure, the ring would be deformed with longer single bonds than double bonds (*Fig. 9b*).

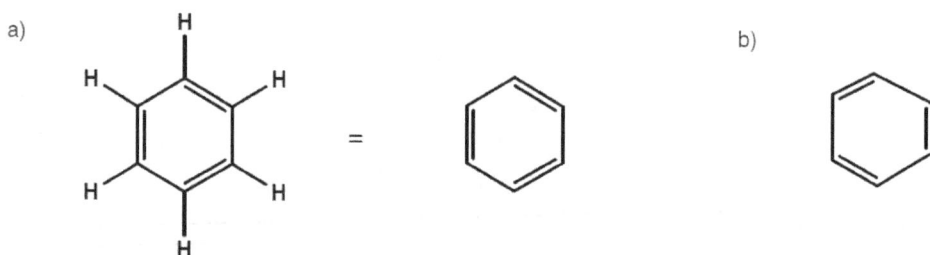

a)   b)

Fig. 9. (a) Representation of the aromatic ring; (b) 'deformed' structure resulting from fixed bonds.

In fact, the C–C bonds in benzene are all the same length. In order to understand this, we need to look more closely at the bonding which takes place. *Figure 10a* shows benzene with all its σ bonds and is drawn such that we are looking into the plane of the benzene ring. Since all the carbons are *sp2* hybridized, there is a *2p* orbital left over on each carbon which can overlap with a *2py* orbital on either side of it (*Fig. 10b*). From this, it is clear that each *2py* orbital can overlap with its neigh- bors right round the ring. This leads to a molecular orbital which involves all the *2py* orbitals where the upper and lower lobes merge to give two doughnut-like lobes above and below the plane of the ring (*Fig. 11a*). The molecular orbital is symmetrical and the six π electrons are said to be delocalized around the aromatic ring since they are not localized between any two particular carbon atoms. The aromatic ring is often represented as shown in *Fig. 11b* to represent this delocalization of the π electrons. Delocalization increases the stability of aromatic rings such that they are less reactive than alkenes (i.e. it requires more energy to disrupt the delocalized π system of an aromatic ring than it does to break the isolated π bond of an alkene).

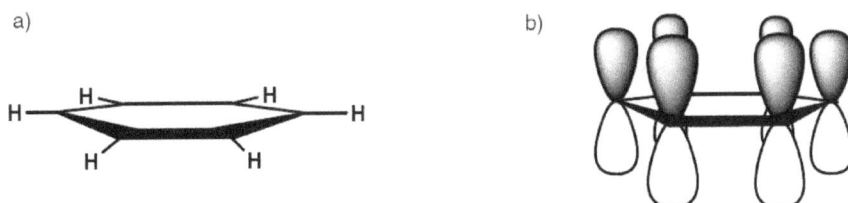

Fig. 10. (a) σ Bonding diagram for benzene, (b) π Bonding diagram for benzene.

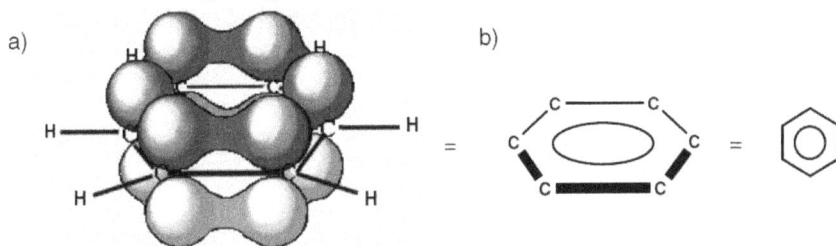

Fig. 11. Bonding molecular orbital for benzene; (b) representation of benzene to illustrate delocalization.

Conjugated systems:

Aromatic rings are not the only structures where delocalization of π electrons can take place. Delocalization occurs in conjugated systems where there are alternating single and double bonds (e.g. 1,3-butadiene). All four carbons in 1,3-butadiene are *sp2* hybridized and so each of these carbons has a half-filled *p* orbital which can interact to give two π bonds (*Fig. 12a*). However, a certain amount of overlap is also possible between the *p* orbitals of the middle two carbon atoms and so the bond connecting the two alkenes has some double bond character (*Fig. 12b*) – borne out by the observation that this bond is shorter in length than a typical single bond. This delocalization also results in increased stability. However, it is important to realize that the conjugation in a conjugated alkene is not as great as in the aromatic system. In the latter system, the π electrons are completely delocalized round the ring and all the bonds are equal in length. In 1,3-butadiene, the π electrons are not fully delocalized and are more likely to be found in the ter- minal C–C bonds. Although there is a certain amount of π character in the middle bond, the latter is more like a single bond than a double bond. Other examples of conjugated systems include α,β-unsaturated ketones and α,β- unsaturated esters (*Fig. 13*). These too have increased stability due to conjugation.

a)

1,3-Butadiene

b)

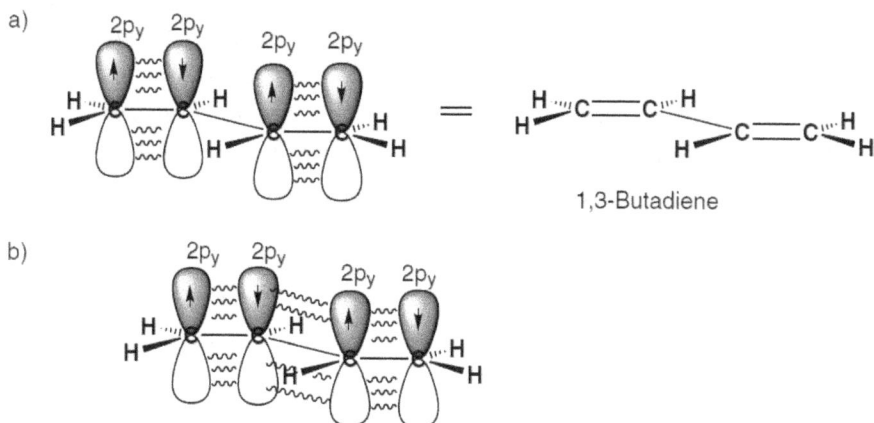

Fig. 12. (a) π Bonding in 1,3-butadiene; (b) delocalization in 1,3-butadiene.

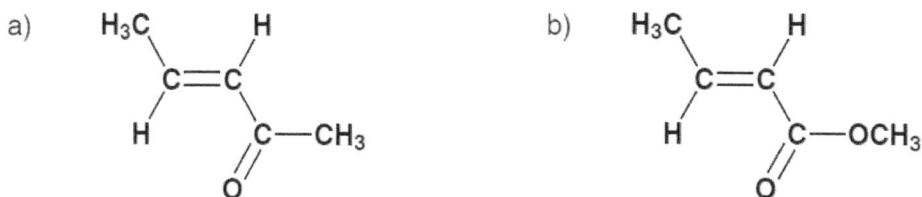

a)

b)

Fig. 13. (a) α,β-Unsaturated ketone; (b) α,β-unsaturated ester.

5.2.3 sp HYBRIDIZATION

Definition:

In *sp* hybridization, the 2s orbital is mixed with one of the 2p orbitals (e.g. 2px) to give two *sp* hybrid orbitals of equal energy. This leaves two 2p orbitals unaffected (2py and 2pz) with slightly higher energy than the hybridized orbitals (*Fig. 1*).

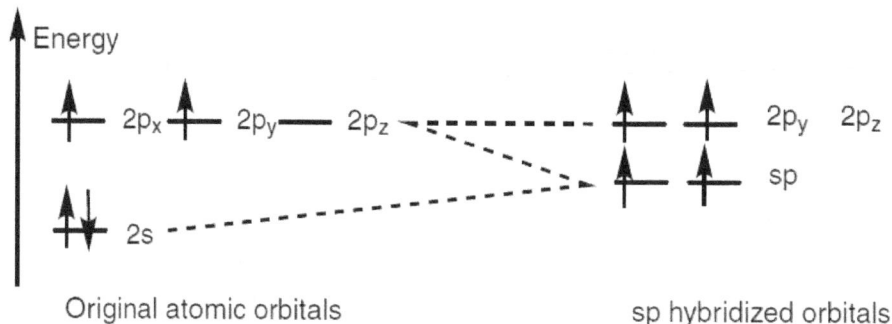

Fig. 1. sp Hybridization of carbon.

Electronic configuration

For carbon, the first two electrons fit into each *sp* orbital according to Hund's rule such that each orbital has a single unpaired electron. This leaves two electrons which can be paired up in the half-filled *sp* orbitals or placed in the vacant 2py and

2pz orbitals. The energy difference between the orbitals is small and so it is easier for the electrons to fit into the higher energy orbitals than to pair up. This leads to two half-filled *sp* orbitals and two half-filled *2p* orbitals (*Fig. 1*), and so four bonds are possible.

Geometry

The *2p* orbitals are dumbbell in shape while the *sp* hybridized orbitals are deformed dumbbells with one lobe much larger than the other. The *2py* and *2pz* orbitals are at right angles to each other (*Fig. 2a*). The *sp* hybridized orbitals occupy the space left over and are in the *x* axis pointing in opposite directions (only the major lobe of the *sp* orbitals are shown in black; *Fig. 2b*).

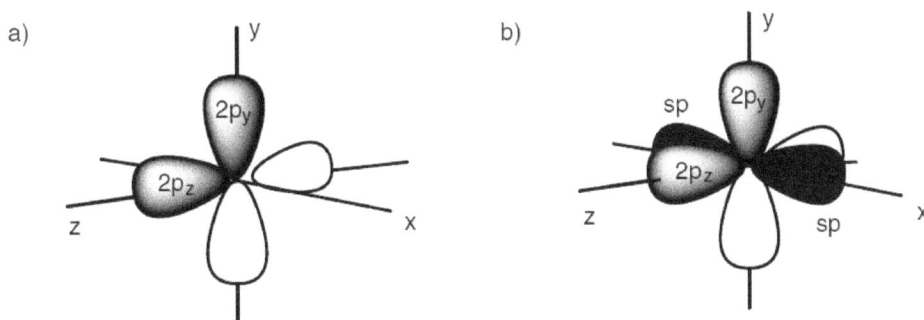

Fig. 2. (a) 2py and 2pz orbitals of an sp hybridized carbon; (b) 2py, 2pz and sp hybridized orbitals of an sp hybridized carbon.

A molecule using the two *sp* orbitals for bonding will be linear in shape. There are two common functional groups where such bonding takes place – alkynes and nitriles.

Alkynes

Let us consider the bonding in ethyne (*Fig. 3*) where each carbon is *sp* hybridized. The C–H bonds are strong σ bonds where each hydrogen atom uses its half-filled

Fig. 3. Ethyne.

1s orbital to bond with a half-filled *sp* orbital on carbon. The remaining *sp* orbital on each carbon is used to form a strong σ carbon–carbon bond. The full σ bonding diagram for ethyne is linear (*Fig. 4a*) and can be simplified as shown (*Fig. 4b*).

Fig. 4. (a) σ Bonding for ethyne; (b) representation of σ bonding.

Further bonding is possible since each carbon has half-filled p orbitals. Thus, the 2py and 2pz orbitals of each carbon atom can overlap side-on to form two π bonds (*Fig. 5*). The π bond formed by the overlap of the 2py orbitals is represented in dark gray. The π bond resulting from the overlap of the 2pz orbitals is represented in light gray. Alkynes are linear molecules and are reactive due to the relatively weak π bonds.

Nitrile groups:

Exactly the same theory can be used to explain the bonding within a nitrile group (C≡N) where both the carbon and the nitrogen are sp hybridized. The energy level diagram in *Fig. 6* shows how the valence electrons of nitrogen are arranged after sp hybridization. A lone pair of electrons occupies one of the sp orbitals, but the other sp orbital can be used for a strong σ bond. The 2py and 2pz orbitals can be used for two π bonds. *Figure 7* represents the σ bonds of HCN as lines and how the remaining 2p orbitals are used to form two π bonds.

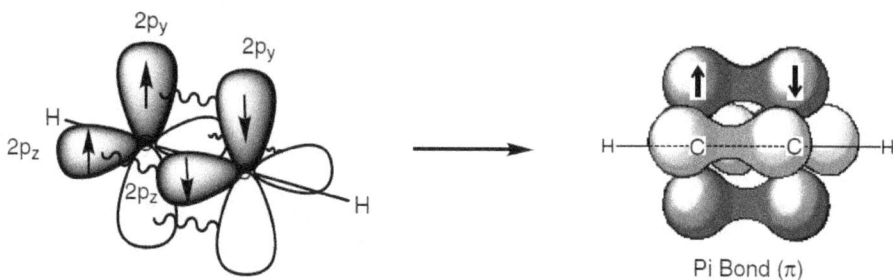

Fig. 5. π-Bonding in ethyne.

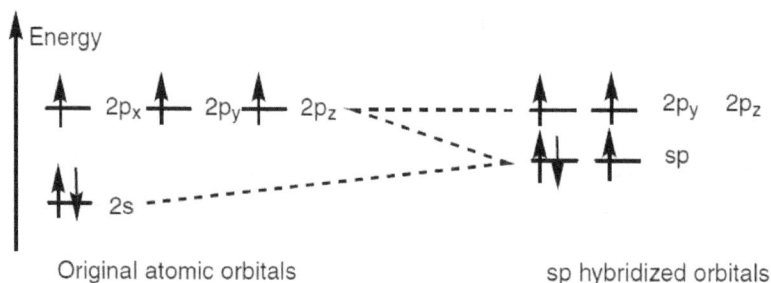

Fig. 6. sp Hybridization of nitrogen.

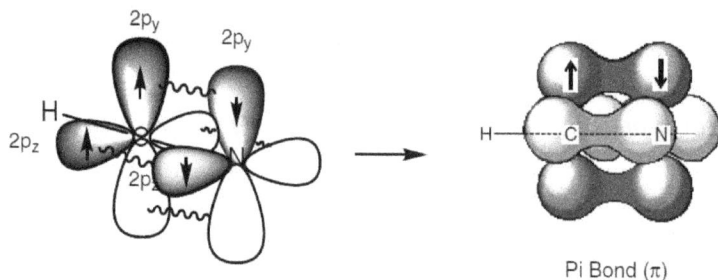

Fig. 7. π -Bonding in HCN.

Bonds And Hybridized Centers

σ and π bonds Identifying σ and π bonds in a molecule (Fig. 1) is quite easy as long as you remember the following rules:

- all bonds in organic structures are either sigma (σ) or pi (π) bonds;
- all single bonds are σ bonds;
- all double bonds are made up of one σ bond and one π bond;
- all triple bonds are made up of one σ bond and two π bonds.

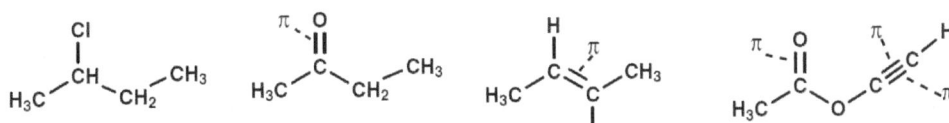

Fig. 1. Examples – all the bonds shown are σ bonds except those labelled as 'IT.

Hybridized centers

All the atoms in an organic structure (except hydrogen) are either *sp*, *sp2* or *sp3* hybridized (*Fig. 2*).

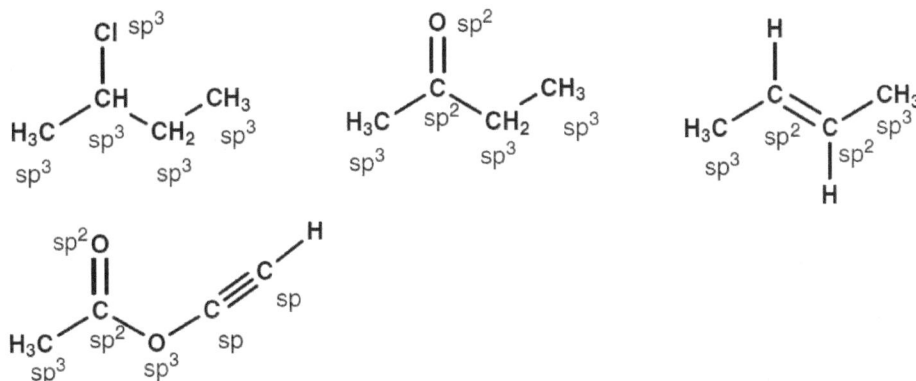

Fig. 2. Examples of sp, sp2 and sp3 hybridized centers.

The identification of sp, sp2 and sp3 centers is simple if you remember the following rules:

- all atoms linked by a single bond are sp3 hybridized (except hydrogen).
- both carbon atoms involved in the double bond of an alkene (C=C) must be sp² hybridized.*
- both the carbon and the oxygen of a carbonyl group (C=O) must be sp2 hybridized
- all aromatic carbons must be sp2 hybridized.
- both atoms involved in a triple bond must be sp hybridized.
- hydrogen uses a 1s orbital for bonding and is not hybridized.

Hydrogen atoms cannot be hybridized. They can only bond by using an *s* orbital since there are no *p* orbitals in the first electron shell. It is therefore impossible for a hydrogen to take part in π bonding. Oxygen, nitrogen and halogens on the other hand can form hybridized orbitals which are either involved in bonding or in holding lone pairs of electrons.

Shape

The shape of organic molecules and the functional groups within them is determined by the hybridization of the atoms present. For example, functional groups containing trigonal planar *sp2* centers are planar while functional groups containing *sp* centers are linear:

- ⊙ Planar functional groups – aldehyde, ketone, alkene, carboxylic acid, acid chloride, acid anhydride, ester, amide, aromatic.
- ⊙ Linear functional groups – alkyne, nitrile.
- ⊙ Functional groups with tetrahedral carbons – alcohol, ether, alkyl halide.

Reactivity

Functional groups which contain π bonds are reactive since the π bond is weaker than a σ bond and can be broken more easily. Common functional groups which contain π bonds are aromatic rings, alkenes, alkynes, aldehydes, ketones, carboxylic acids, esters, amides, acid chlorides, acid anhydrides, and nitriles.

  *Functional groups known as allenes (R2C=C=CR2) have an *sp* hybridized carbon located at the center of two double bonds, but these functional groups are beyond the scope of this text.

5.3 MOLECULAR GEOMETRY AND VESPER THEORY

Understanding molecular geometry is a cornerstone of modern chemistry, enabling chemists to predict and rationalize the three-dimensional arrangements of atoms within molecules. At the heart of molecular geometry lies the Valence Shell Electron Pair Repulsion (VSEPR) theory, which provides a simple yet powerful framework for predicting molecular shapes based on the arrangement of electron pairs around a central atom. The VSEPR theory states that electron pairs in the valence shell of an atom repel each other and therefore adopt positions in space that minimize repulsion, resulting in specific molecular geometries. These geometries profoundly influence a molecule's physical and chemical properties, including its reactivity, polarity, and biological activity. By applying the principles of VSEPR theory, chemists can accurately predict molecular shapes, understand molecular interactions, and design molecules with tailored properties for applications ranging from drug design to materials science. Thus, the study of molecular geometry and VSEPR theory stands as a cornerstone in the quest to unravel the mysteries of chemical structure and behavior, driving innovations across diverse fields of science and technology.

Molecules have shapes. There is an abundance of experimental evidence to that effect—from their physical properties to their chemical reactivity. Small molecules—molecules with a single central atom—have shapes that can be easily predicted. The basic idea in molecular shapes is called valence shell electron pair repulsion (VSEPR). It basically says that electron pairs, being composed of negatively charged particles, repel each other to get as far away from each other as possible. VSEPR makes a distinction between electron group geometry, which expresses how electron groups (bonds and nonbonding electron pairs) are arranged, and molecular geometry, which expresses how the atoms in a molecule are arranged. However, the two geometries are related.

There are two types of electron groups: any type of bond—single, double, or triple—and lone electron pairs. When applying VSEPR to simple molecules, the first thing to do is to count the number of electron groups around the central atom. Remember that a multiple bond counts as only one electron group.

Any molecule with only two atoms is linear. A molecule whose central atom contains only two electron groups orients those two groups as far apart from each other as possible—180° apart. When the two electron groups are 180° apart, the atoms attached to those electron groups are also 180° apart, so the overall molecular shape is linear. Examples include BeH_2 and CO_2:

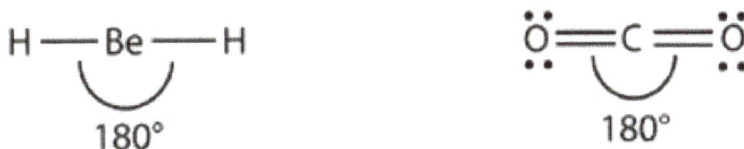

Figure 1 Beryllium hydride and carbon dioxide bonding.

The two molecules, shown in the figure below in a "ball and stick" model.

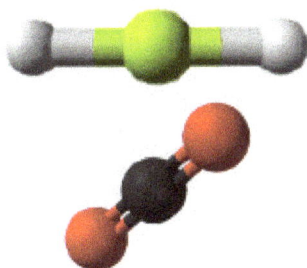

Figure 2 Beryllium hydride and carbon dioxide models.

A molecule with three electron groups orients the three groups as far apart as possible. They adopt the positions of an equilateral triangle—120° apart and in a plane. The shape of such molecules is trigonal planar. An example is BF_3:

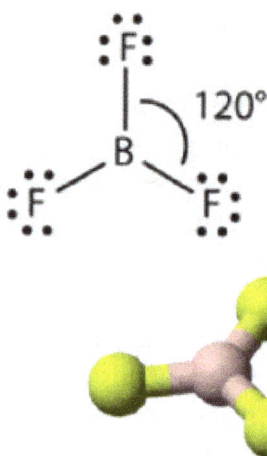

Figure .3 Boron trifluoride bonding.

Some substances have a trigonal planar electron group distribution but have atoms bonded to only two of the three electron groups. An example is GeF_2:

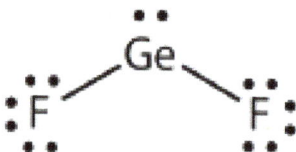

Figure 4: Germanium difluoride bonding.

From an electron group geometry perspective, GeF_2 has a trigonal planar shape, but its real shape is dictated by the positions of the atoms. This shape is called *bent* or *angular*.

A molecule with four electron groups about the central atom orients the four groups in the direction of a tetrahedro. If there are four atoms attached to these electron groups, then the molecular shape is also *tetrahedral*. Methane (CH_4) is an example.

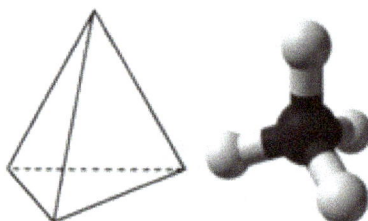

Figure 5 Tetrahedral structure of methane.

This diagram of CH_4 illustrates the standard convention of displaying a three-dimensional molecule on a two-dimensional surface. The straight lines are in the plane of the page, the solid wedged line is coming out of the plane toward the reader, and the dashed wedged line is going out of the plane away from the reader.

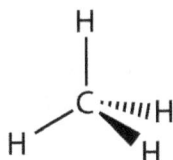
Figure 6 Methane bonding.

NH_3 is an example of a molecule whose central atom has four electron groups but only three of them are bonded to surrounding atoms.

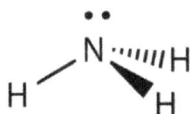
Figure 7 Ammonia bonding.

Although the electron groups are oriented in the shape of a tetrahedron, from a molecular geometry perspective, the shape of NH_3 is *trigonal pyramidal*.

H_2O is an example of a molecule whose central atom has four electron groups but only two of them are bonded to surrounding atoms.

Figure 8 Water bonding.

Although the electron groups are oriented in the shape of a tetrahedron, the shape of the molecule is *bent* or *angular*. A molecule with four electron groups about the central atom but only one electron group bonded to another atom is linear because there are only two atoms in the molecule.

Double or triple bonds count as a single electron group. The Lewis electron dot diagram of formaldehyde (CH_2O) is shown in Figure 9.

Figure .9 Lewis electron dot diagram of formaldehyde.

The central C atom has three electron groups around it because the double bond counts as one electron group. The three electron groups repel each other to adopt a trigonal planar shape.

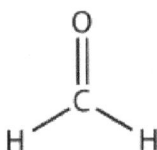
Figure .10: Formaldehyde bonding.

(The lone electron pairs on the O atom are omitted for clarity.) The molecule will not be a perfect equilateral triangle because the C–O double bond is different from the two C–H bonds, but both planar and triangular describe the appropriate approximate shape of this molecule. Figure 11 illustrates several representations of the water, ammonia, and methane molecules.

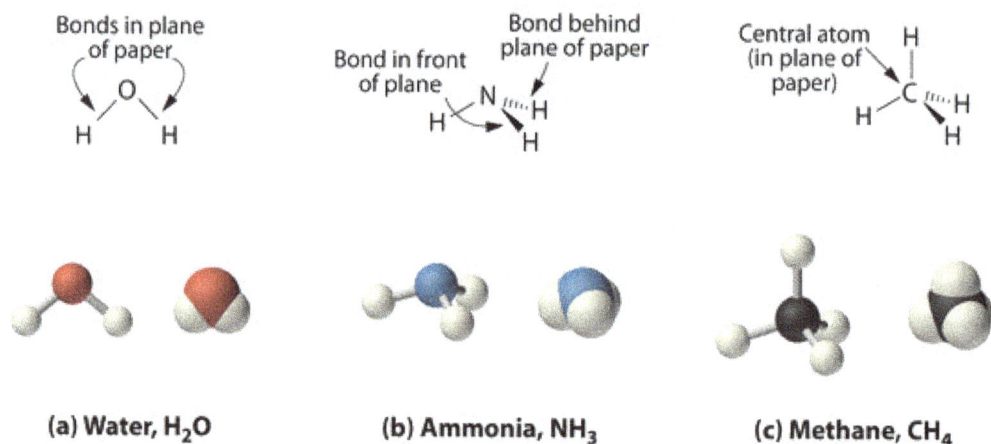

(a) Water, H$_2$O   (b) Ammonia, NH$_3$ (c) Methane, CH$_4$

Figure 11. The three-dimensional structures, ball and stick models, and space filling models of water, ammonia, and methane. (a) Water is a V-shaped molecule, in which all three atoms lie in a plane. (b) In contrast, ammonia has a pyramidal structure, in which the three hydrogen atoms form the base of the pyramid and the nitrogen atom is at the vertex. (c) The four hydrogen atoms of methane form a tetrahedron; the carbon atom lies in the center.

Table .1: Summary of Molecular Shapes

Number of Electron Groups on Central Atom	Number of Bonding Groups	Number of Lone Pairs	Electron Geometry	Molecular Shape
2	2	0	linear	linear
3	3	0	trigonal planar	trigonal planar
3	2	1	trigonal planar	bent
4	4	0	tetrahedral	tetrahedral
4	3	1	tetrahedral	trigonal pyramidal
4	2	2	tetrahedral	bent

5.4 MOLECULAR ORBITAL THEORY

Lewis dot, VESPR & Valence Bond (VB) theories all do a good job at predicting the shapes and bonding in covalent molecules.

Chemists however sometimes require another theory of bonding that explains phenomenon such as *paramagnetism* in covalent molecules or why certain species form unstable compounds while others don't.

To address such issues, we can describe bonding from a perspective of **Molecular Orbitals**. In MO theory, the **atomic orbitals** on individual atoms combine (*constructively and destructively*) to produce new **molecular orbitals** that give rise to bonding and anti-bonding orbitals that exist between nuclei.

Molecular Orbital Theory has several advantages and differences over VESPR & VB theory:

◉ MO does a good job of predicting electronic spectra and paramagnetism, when VSEPR and the VB theories don't.

◉ The MO theory like VB theory, predicts the bond order of molecules, however it does not need resonance structures to describe molecules

◉ MO theory treats molecular bonds as a sharing of electrons between nuclei. Unlike the VB theory, which treats the electrons as **localized hybrid orbitals** of electron density.

◉ MO theory says that the electrons are **delocalized**. That means that they are spread out over the entire molecule.

◉ The main drawback to our discussion of MO theory is that we are limited to talking about **diatomic molecules** (molecules that have only two atoms bonded together), or the theory gets very complex.

When two atoms come together, their atomic orbitals interact to form two possible molecular orbitals, (1) the lower energy "**bonding**" MO and (2) the higher energy "**anti-bonding**" MO.

Recall that the atomic orbitals on an atom are described by **Wave Functions (Y)**, *i.e.* the mathematical representations of the probability of space in which an electron resides.

Considered a molecule A-B

When two wave functions (orbitals) on different atoms add constructively they produce a new MO that promotes boding given by:

(1) YA + YB à WAB

When two wave functions (orbitals) on different atoms add destructively (subtract) they produce a new MO that promotes anti-bonding boding given by:

(2) YA - YB à W*AB

◉ The lower energy bonding molecular orbital (1) stabilizes the molecule is.

◉ The higher energy anti-bonding orbital (2) destabilizes the molecule.

Consider H_2

When the 1s orbitals of each H-atom subtract, an anti-bonding orbital forms.

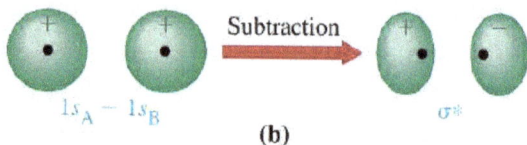

Subtraction

$1s_A - 1s_B$

(b)

$\sigma*$

When the 1s orbitals of each H-atom add, a bonding orbital forms.

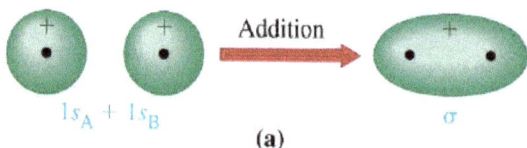

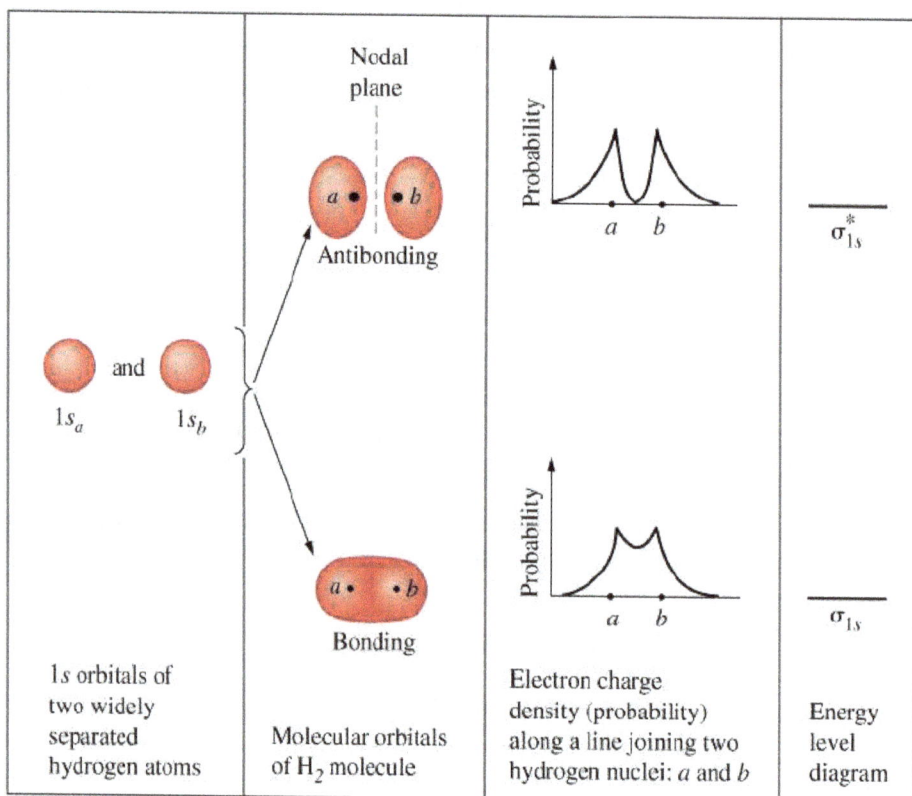

Addition

$1s_A + 1s_B$

(a)

σ

Energy

Nodal plane

$a \quad b$

Antibonding

Probability

$a \quad b$

σ_{1s}^*

$1s_a$ and $1s_b$

$a \quad b$

Bonding

Probability

$a \quad b$

σ_{1s}

| 1s orbitals of two widely separated hydrogen atoms | Molecular orbitals of H_2 molecule | Electron charge density (probability) along a line joining two hydrogen nuclei: *a* and *b* | Energy level diagram |

5.4.1 Terminology in MO theory:

⊙ *σ-bonding orbital*: stabilizing MO that exists between nuclei.

⊙ *σ*-anti-bonding orbital*: destabilizing MO that exists between nuclei.

⊙ *π-bonding orbital*: stabilizing MO that exists above and below the inter nuclear axis.

⊙ *π *-anti-bonding orbital*: destabilizing MO that exists above and below the inter nuclear axis.

- ⊙ **node**: region of zero probability
- ⊙ **ground state**: lowest energy configuration of electrons in a molecule
- ⊙ **excited state**: electron(s) moved to higher energy MO's
- ⊙ **Bond Order**: $BO = \dfrac{\text{\# of bonding electrons in MO's - \# of anti - bonding electrons in MO's}}{2}$
- ⊙ **LUMO**: lowest energy occupied MO
- ⊙ **HOMO**: highest energy occupied MO

5.4.2 MO Diagrams:

Once again we consider the simplest molecule, H2. When two hydrogen atoms combine the 1s orbitals on each can add or subtract to form a s(1s) bonding or s*(1s) antibonding orbital.

Each H-atom has one 1s electron that can contribute to the MO bonding and anti-bonding MO's. Just as in the electron configurations of atoms, the electron fill from the lowest energy MO first (**Aufbau principle**) only pairing when forced to (**Hund's rule**). Each MO can only hold two electrons of opposite spin (**Pauli priciple**)

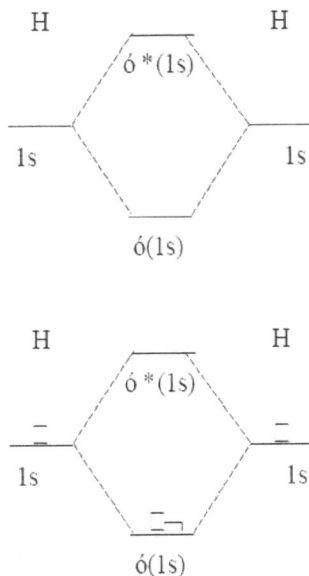

MO Diagrams: From this we find that the bonding in H2 is described by: Bond Order = 1 (single bond)

$$BO = \frac{\text{\# of bonding electrons in MO's - \# of anti - bonding electrons in MO's}}{2} = \frac{2-0}{2} = 1$$

MO electron configuration of: $ó_{1s}^{2}$

And we conclude that the molecule (H2) is **diamagnetic** (no unpaired electrons)

MO Diagrams: He2

H2, O2, N2 and the halogens all exist in nature as stable diatomic molecules.

Why is it then that He is a monatomic species?

Let's use MO theory to explain:

- ⊙ Each He-atom has two 1s electron that can contribute to the MO bonding and anti-bonding MO's.

- ⊙ To determine whether He2 is a stable molecule, we fill the MO diagram for the 1s system just like H2.

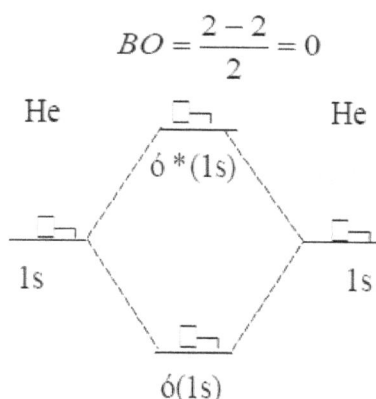

$$BO = \frac{2-2}{2} = 0$$

He   ⁻ᴦ   He

ó*(1s)

‾ᴦ   ‾ᴦ

1s   1s

⁻ᴦ

ó(1s)

From this we see that He2 is an unstable molecule. The bond order is **zero**!

MO theory recap:

- ⊙ The number of **molecular orbitals** (MO's) formed is equal to the number of **atomic orbitals** (AO's) that combine.

- ⊙ When AO's combine they **add** (constructive) and **subtract** (destructive) to form bonding and anti-bonding MO's respectively

- ⊙ The bonding MO's are always at a lower state of energy than the anti-bonding MO's.

- ⊙ The maximum number of electrons in any MO is two.

- ⊙ The electrons from each atom that form the MO's fill from the lowest energy states first, pairing only when forced to. This is the **ground state**.

- ⊙ When electrons in the ground state are promoted to higher energy MO's, an **excited state** results.

Molecular Orbitals and the p Atomic Orbitals

Recall that the p-orbitals of an atom are oriented along the axes of a Cartesian set of coordinates. Each lobe is orthogonal to one another (90°).

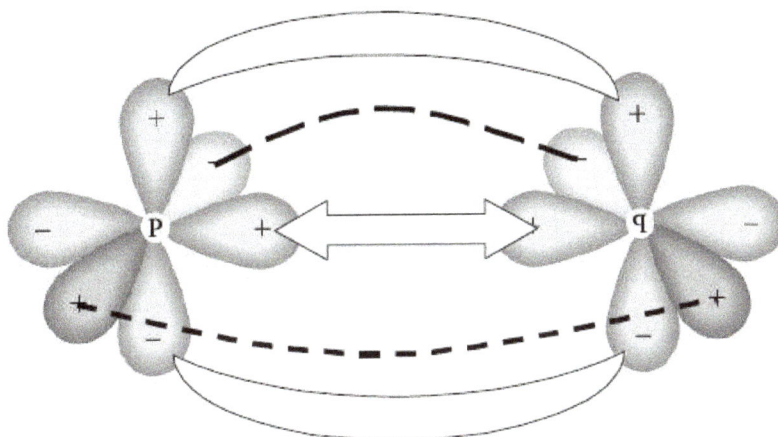

- When the lobes add and subtract, three sets of bonding and three sets of antibonding MO's result.
- One has sigma character (s) and two have pi (p) character. One addition/ subtraction results in (a) & (b), s(2p) and s*(2p)

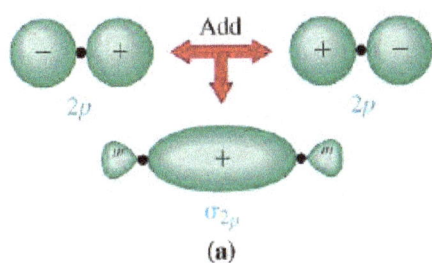

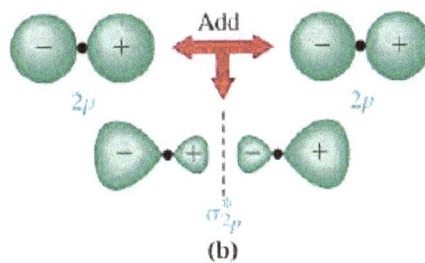

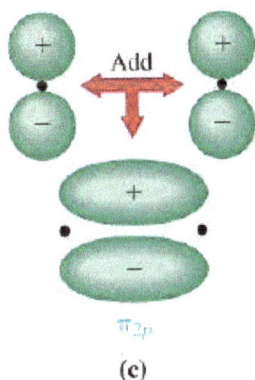

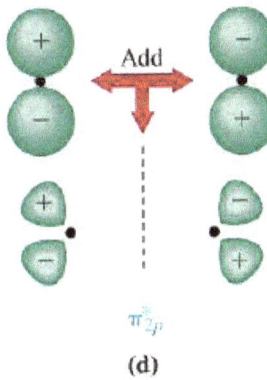

The other results in two degenerate p(2p) and p*(2p) MO's.

End on produces the
sigma system.

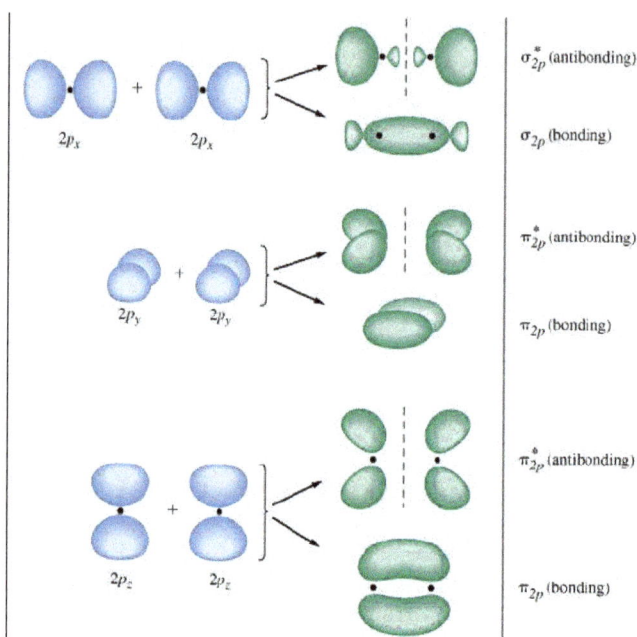

$2p_x$ $2p_x$ σ_{2p}^{*} (antibonding)

σ_{2p} (bonding)

Above and below
produced the two pi
systems.

$2p_y$ $2p_y$ π_{2p}^{*} (antibonding)

π_{2p} (bonding)

$2p_z$ $2p_z$ π_{2p}^{*} (antibonding)

π_{2p} (bonding)

The p(2p) System (Z_7):

The π(2p) System (Z$^-$7):

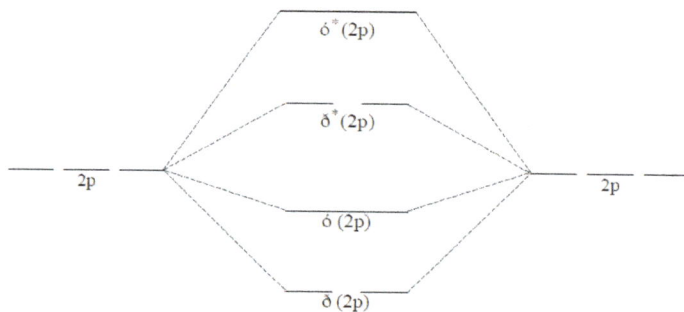

ó* (2p)

δ* (2p)

2p   2p

ó (2p)

δ (2p)

The p(2p) System (Z³8):

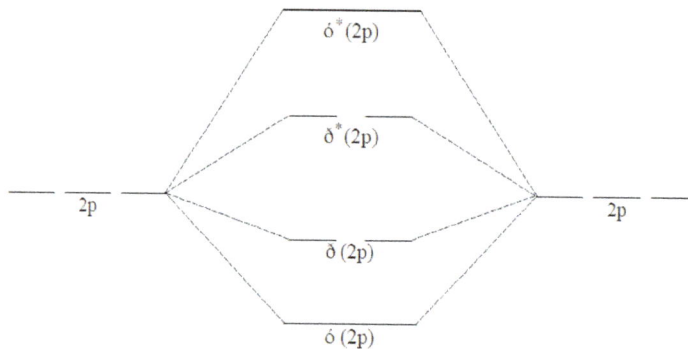

ó* (2p)

δ* (2p)

2p   2p

δ (2p)

ó (2p)

06 | MOLECULAR FORCES AND INTERMOLECULAR BONDING

6.1 DIPOLE-DIPOLE FORCES

Dipole-dipole forces are the attractive forces that occur between polar molecules. A molecule of hydrogen chloride has a partially positive hydrogen atom and a partially negative chlorine atom. In a collection of many hydrogen chloride molecules, the molecules will align themselves so that the oppositely charged regions of neighboring molecules are near each other.

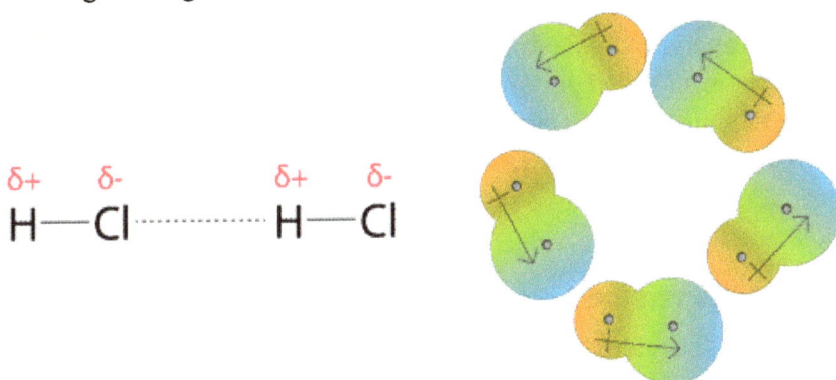

Figure 1.: Dipole-dipole forces are a result of the attraction of the positive end of one dipole to the negative end of a neighboring dipole.

Dipole-dipole forces are similar in nature to ionic bonds, but much weaker.

6.2 LONDON DISPERSION FORCES

Dispersion forces are the weakest of all intermolecular forces. They are often called London dispersion forces after Fritz London (1900-1954), who first proposed their existence in 1930. London dispersion forces are the intermolecular forces that occur between atoms, and between nonpolar molecules as a result of the motion of electrons.

The electron cloud of a helium atom contains two electrons, which can normally be expected to be equally distributed spatially around the nucleus. However, at any given moment the electron distribution may be uneven, resulting in an instantaneous dipole. This weak and temporary dipole subsequently influences neighboring helium atoms through electrostatic attraction and repulsion. It induces a dipole on nearby helium atoms.

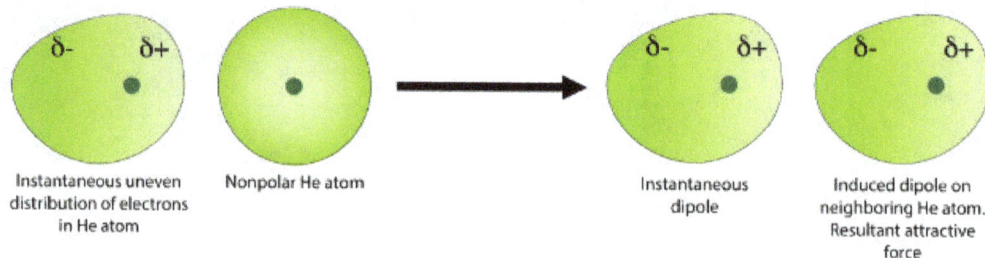

Instantaneous uneven distribution of electrons in He atom

Nonpolar He atom

Instantaneous dipole

Induced dipole on neighboring He atom. Resultant attractive force

Figure 2: A short-lived or instantaneous dipole in a helium atom.

6.3 LONDON DISPERISON FORCES

The atoms are combined to form molecules. In a molecule, atoms are bonded with chemical bonds. Chemical bonds are formed by sharing electrons between atoms. On the basis of sharing of electrons between atoms, chemical bonds can be classified in different types such as ionic, covalent, metallic and coordination bonds.

The instantaneous dipole–induced dipole attractions are called London dispersion forces after Fritz London (1900–1954), a German physicist who developed this model to explain the intermolecular attractions that exist between non- polar molecules. London's dispersion forces occur between all molecules. These very weak attractions occur because of the random motions of electrons on atoms within molecules.

6.3.1 London Dispersion Forces Definition

So, we can say that covalent bond, ionic bond, and coordination bond are the intra-molecular force of attraction which form within a molecule. The forces of attraction between molecules which hold them together are called the intermolecular force of attraction. These forces are weaker than intermolecular forces. These forces are responsible for the liquids, solids and solutions state of any compound. Some common types of intermolecular forces are London dispersion, dipole-dipole, Hydrogen bonding and ion-ion force.

The order of strength of these intermolecular forces is given below.

London's dispersion force < dipole-dipole < H-bonding < Ion-ion

So, we can say that London dispersion forces are the weakest intermolecular force. London's dispersion forces can be defined as a temporary attractive force due to the formation of temporary dipoles in a nonpolar molecule.

When the electrons in two adjacent atoms are displaced in such a way that atoms get some temporary dipoles, they attract each other through the London dispersion force. These intermolecular forces occur between non-polar substances. Due to these forces, they can condense to liquids and or freeze into solids at low temperature.

6.3.2 Types of Bonds

Ionic Bonds

Ionic bonds are formed by the formation of cation and anions. An atom forms cation after losing of electron and such ions has a positive charge. If an atom accepts electrons, it results in the formation of anion which has a negative charge. Cation and anion attract each other to form an ionic bond. So, we can say that ionic bonds are an electrostatic force of attraction between oppositely charged ions.

For example; NaCl is an ionic compound in which Na+ and Cl– combine to form an ionic compound; sodium chloride. Covalent bonds are formed by equal sharing of electrons between bonded atoms. All atoms tend to complete the octet configuration that provides stability to them.

Covalent bonds

The sharing of electrons helps to get the octet configuration to both bonded atoms. Covalent bonds are usually formed between two non-metals. They can be polar or nonpolar in nature. The polarity of covalent bonds depends on the electronegativity of both bonded atoms. We know that metals have a tendency to lose electrons and form metal cations. These free mobile electrons remain in the metallic lattice. The electrostatic force of attraction between metal ions and free mobile electrons is called a metallic bond. The unique physical properties of metals such as malleability, ductility etc are due to this metallic bond only.

Coordination bonds

Coordination bonds are basically a type of covalent bond which is formed by unequal sharing of electrons between two atoms. Here one atom acts as acceptor and other acts as a donor. These chemical bonds are formed between atoms to form molecules. There are several attraction forces between molecules like dipole-dipole interaction, dipole- induced dipole interaction, Vander Wall interaction, Hydrogen bonding, London dispersion forces etc.

6.3.3 London Dispersion Forces Example

The unequal distribution of electrons about the nucleus in an atom can induce some dipole in the atom. When another atom or molecule comes in contact with this induced dipole, it can be distorted that leads to an electrostatic attraction between either atoms or molecules.

LONDON DISPERSION FORCES

BYJU'S
The Learning App

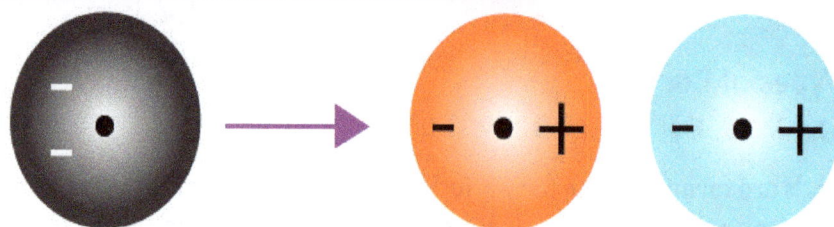

Uneven distribution of electrons in He

Instantaneous dipole

Induced dipole on neighboring He

If these atoms or molecules touch each other, dispersion forces are present between any of them.

For example, consider London dispersion forces between two chlorine molecules. Here both chlorine atoms are bonded through a covalent bond which forms by equal sharing of valence electrons between two chlorine atoms. The force of attraction between two chlorine molecules is the London dispersion force here which is due to unequal distribution of electron density in the molecule.

6.3.4 London Dispersion Forces Formula

The tendency of molecules to form charge separation or induced dipole is called polarizability. The interaction between two dipoles can be expressed as its strength which is denoted as μ. The strength is directly proportional to the strength of the electric field (E).

$$\mu = \alpha \times E$$

Here,

μ = Induced dipole moment

α = Polarizability

E = Electric field

The interaction energy can be calculated with the help of London dispersion force formula.

$$V_{11} = -\frac{3\alpha^2 I}{4r^6}$$

This formula is for the potential energy between two identical atoms or molecules. The formula was modified by German physicist, Fritz London for two un-identical atoms or molecules as given below.

$$V_{12} = -\frac{3I_1 I_2 a_1' a_2'}{2I_1 + I_2 r^6}$$

Here

I = Ionization energy

A = Polarizability

r = Distance between molecules

6.3.5 London Dispersion Forces vs Van der Waals Forces

- In general, all the intermolecular forces of attraction between molecules are called Van der Waals forces.

- Van der Waals forces can be classified as weak London dispersion Forces and stronger dipole-dipole forces.

- Both of these forces are due to momentarily dipole formation. The displacement of electrons causes a nonpolar molecule to be a polar molecule.

- The capability of a molecule to become polar is called polarizability of molecules. As we move from top to bottom in a group of the periodic table, the polarizability increases whereas it increases from right to left within periods.

- As the polarity in the molecule increases, the melting and boiling points also increase as more heat is needed to break the bonds.

- So we can say that as the mass increases, the number of electrons increases, and melting and boiling points also increase. Long-chain molecules exhibit strong London dispersion forces because more displacement can be possible in such molecules.

- London Dispersion Forces vs Dipole DipoleBoth London dispersion forces and dipole-dipole interactions are types of Van der Wall forces.

- London's dispersion forces are weaker than dipole-dipole forces as they are because of momentarily dipoles.

- The dipole-dipole interactions are due to interaction of partially positively charged a part of a molecule with the partially negatively charged part of the neighbouring molecule.

- So we can say that these interactions are in polar molecules such as water, hydrochloric acid etc.

6.4 COMPARISION OF INTERMOLECULAR FORCES

Intermolecular forces are the attractive forces that exist between molecules. They play a crucial role in determining the physical properties of substances, such as melting point, boiling point, viscosity, and solubility. The strength of intermolecular forces varies depending on the types of molecules involved. Here's a comparison of the main types of intermolecular forces:

6.4.1 Van der Waals Forces:

⊙ *London Dispersion Forces:* These are the weakest intermolecular forces and occur between all molecules, regardless of their polarity. They arise due to temporary fluctuations in electron density, which create temporary dipoles in neighboring molecules.

⊙ *Dipole-Dipole Interactions:* These forces occur between polar molecules. They result from the attraction between the positive end of one polar molecule and the negative end of another polar molecule. Dipole-dipole interactions are stronger than London dispersion forces but weaker than hydrogen bonds.

⊙ *Hydrogen Bonds:* Hydrogen bonds are a special type of dipole-dipole interaction that occurs between a hydrogen atom bonded to a highly electronegative atom (such as oxygen, nitrogen, or fluorine) and a lone pair of electrons on another electronegative atom. Hydrogen bonds are stronger than both London dispersion forces and dipole-dipole interactions.

6.4.2 Ionic Interactions:

⊙ Ionic interactions occur between ions of opposite charges. They are the strongest type of intermolecular force and are responsible for the properties of ionic compounds such as salts.

6.4.3 Hydrophobic Interactions:

⊙ Hydrophobic interactions arise between nonpolar molecules in aqueous environments. These interactions are driven by the tendency of nonpolar molecules to minimize contact with polar water molecules.

6.4.4 Ion-Dipole Interactions:

⊙ These interactions occur between an ion and a polar molecule. They are relatively strong and are important in solutions containing ionic compounds dissolved in polar solvents.

In summary, the strength of intermolecular forces follows this general order: Ionic interactions > Hydrogen bonds > Dipole-dipole interactions > London dispersion forces. Understanding the nature and strength of intermolecular forces is essential for explaining many physical and chemical phenomena, including phase transitions, solubility, and molecular structure.

07 | MOLECULAR ORBITAL THEORY

7.1 FORMATION OF MOLECULAR ORBITAL

The formation of molecular orbitals (MOs) involves the combination and interaction of atomic orbitals from individual atoms to create new orbitals that describe the distribution of electrons in a molecule. This process is fundamental in understanding the electronic structure and bonding in molecules, including diatomic molecules. When two atoms come together to form a molecule, their atomic orbitals overlap, leading to the formation of molecular orbitals. The combination of atomic orbitals results in the formation of bonding and antibonding molecular orbitals. Bonding molecular orbitals are characterized by a lower energy compared to the original atomic orbitals, indicating a stabilizing effect on the molecule. In contrast, antibonding molecular orbitals have higher energy levels, making them less stable configurations. The simplest example of molecular orbital formation occurs in diatomic molecules, where atomic orbitals from two atoms combine to form bonding and antibonding molecular orbitals. For instance, in the case of hydrogen (H_2), the 1s atomic orbitals of two hydrogen atoms overlap to produce a bonding σ (sigma) orbital and an antibonding σ* orbital. The bonding σ orbital has a lower energy level and contains a higher electron density between the two nuclei, contributing to the stability of the molecule. Conversely, the antibonding σ* orbital has higher energy and results in electron density away from the internuclear region, destabilizing the molecule. The process of molecular orbital formation follows the principles of quantum mechanics, where the wave functions of atomic orbitals combine to form new wave functions representing the molecular orbitals. The combination of atomic orbitals can occur in various ways, including constructive and destructive interference, resulting in the formation of bonding and antibonding molecular orbitals, respectively. The concept of molecular orbitals provides a powerful framework for understanding chemical bonding, electronic structure, and the properties of molecules. It allows scientists to predict the stability, reactivity, and spectroscopic behaviour of molecules, contributing to advancements in fields such as chemistry, materials science, and molecular biology. Additionally, molecular orbital theory serves as the foundation for computational methods used to study complex molecular systems and design new materials with tailored properties for specific applications.

7.1.1 Molecular Orbitals of H_2

As two H atoms come together their orbitals will overlap, allowing the electrons to move from one atom to the other and *vice versa*. The electrons no longer belong to just one atom, but to the molecule. They are now delocalized over the whole molecule, shared by the atoms.

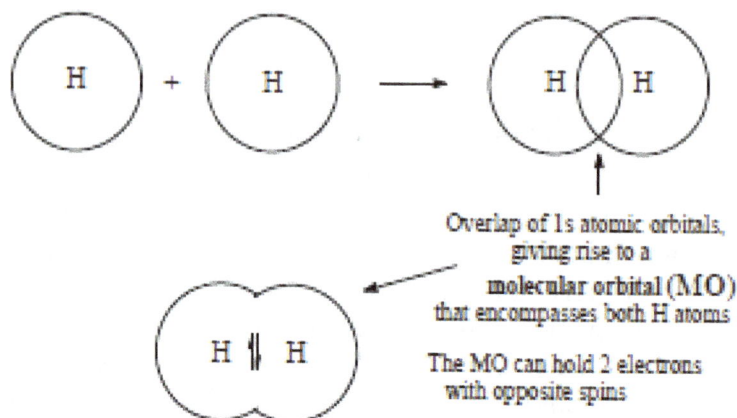

Overlap of 1s atomic orbitals, giving rise to a molecular orbital (MO) that encompasses both H atoms

The MO can hold 2 electrons with opposite spins

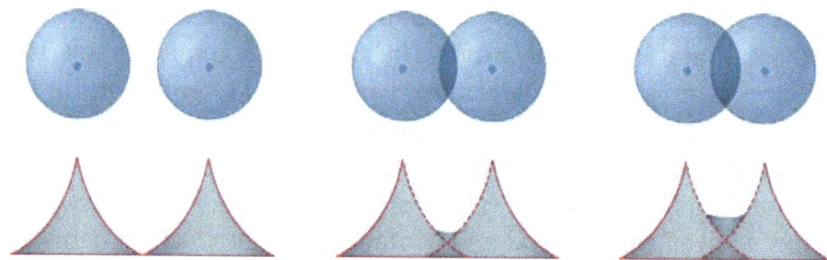

As two hydrogen atoms approach each other, the overlap of their 1s atomic orbitals increases. The wave amplitudes add to generate a new orbital with high electron density between the nuclei.

In general, as

Atoms → Molecule

Atomic orbitals (AO) → Molecular Orbitals (MO) MO's are formed by combining (overlapping) AO's.

Bonding occurs if it's energetically more favourable for the electrons to be in MO's (*i.e.* in a molecule) rather than in AO's (*i.e.* in individual atoms).

7.1.2 Bonding and Antibonding MO's

The combination of two AO's can be in phase →low energy bonding MO out of phase → high energy antibonding MO

out of phase combination (-)   high energy (antibonding) MO
orbital changes sign

+/-

node (where MO changes sign)

in phase combination (+)   low energy (bonding) MO
orbital positive everywhere

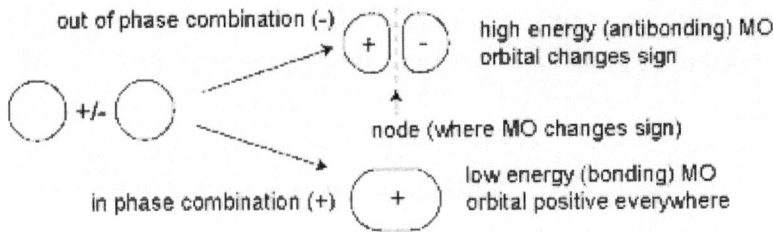

MO energy level diagram:

Now, feed in electrons

H2+ σ1 stable

$(E(H2+) < E(H) + E(H+))$

H2 σ 2 stable

$He_2^+ \sigma^2 \sigma^{*1}$ stable

$He_2 \sigma^2 \sigma^{*2}$ not stable!

$E(He2) \sim 2E(He)$

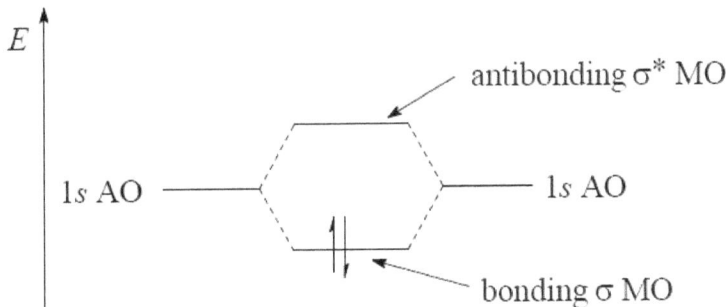

In fact, one electron in the (bonding) MO is sufficient for bonding to occur, *i.e.* H2+ is predicted to be stable! (It has been verified by experiments.) Atoms will be bonded (in a molecule) provided there is an excess of bonding electrons.

H2 , in addition to the lowest σ and σ * MO's, has other higher energy MO's with corresponding allowed energies, formed from the 2s, 2p,...AO's. All these MO's have lobe structures and nodes reminiscent of atomic orbitals.

The energy of the H2 molecule is *lower than* the energy of two isolated H atoms. That is, the energy change associated with bond formation is negative.

We call this molecular orbital a *bonding orbital* for this very reason. It is symmetric to rotation about the interatomic axis, hence it's called a σ MO.

The other orbitals have higher energies than the atomic orbitals of H.

Electrons in these orbitals would not contribute to the stability of the molecule; in fact they would result in destabilization.

H_2 contains the simplest kind of bond, provided by a pair of shared electrons delocalised around two nuclei in a s MO. The bond is therefore known as a sigma (σ) bond.

The next-lowest energy orbital is unoccupied. It lies above the energy of the $1s$ atomic orbitals (from which it's built), hence we refer to it as an *anti-bonding orbital*.

Look also at the shape of the lobes: the *anti-bonding orbital* has a node between the two nuclei.

Where the bonding orbital has an electron density build-up between the nuclei, the anti- bonding orbital would have a reduced electron density ($\Psi 2$).

The solution to the Wave Equation for molecules leads to quantum states with discrete energy levels and well-defined shapes of electron waves (molecular orbitals), *just like atoms*.

Each orbital contains a maximum of two (spin-paired) electrons, *just like atoms*.

"Bonds" form because the energy of the electrons is lower in the molecules than it is in isolated atoms. Stability is conferred by electron *delocalization* in the molecule. This is a quantum effect: the more room an electron has, the lower its (kinetic) energy. Therefore, the existence of molecules is a direct consequence of the quantum nature of electrons.

This gives us a convenient picture of a bond in terms of a pair of shared (delocalized) electrons. It also suggests simple (and commonly-used) ways of representing simple sigma bonds as:

- ◉ A shared pair of electrons (in a bonding MO) H : H
- ◉ A line between nuclei H-H

7.2 BONDING OF MULTI-ELECTRON ATOMS

What kinds of orbitals and bonds form when an atom has more than one electron to share?

We will step up the complexity gradually, first considering other diatomic molecules. These fall into two classes

1. *Homonuclear Diatomics.* These are formed when two *identical* atoms combine to form a bond. E.g. H_2, F_2, Cl_2, O_2...

2. *Heteronuclear Diatomics.* These are formed when two *different* atoms combine to form a bond. E.g. HF, NO, CO, ClBr

A general and systematic approach to the construction of MO's of a homonuclear diatomic molecule is to consider pair-wise interactions between atomic orbitals of the *same energy* and *symmetry*.

Given the $1sa$, $2sa$ and $2pa$ AO's on atom a and $1sb$, $2sb$ and $2pb$ AO's on atom b, we can form the following *bonding* and *antibonding* MO's of σ and σ symmetry:

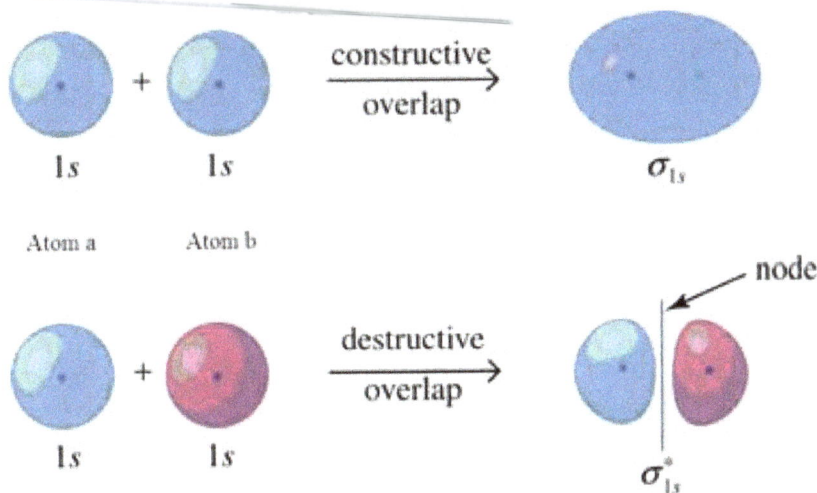

Head to head combination of p type AO's also results in π and π^* MO's

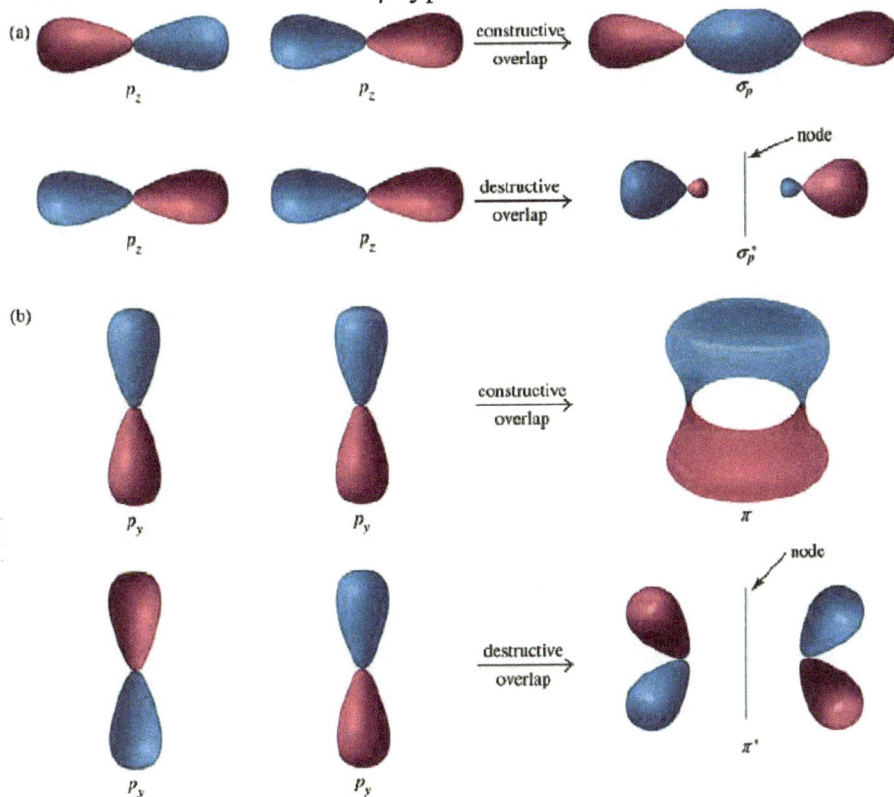

Sideways (parallel) combination of p type AO's results in π and π^* MO's.

As there are two equivalent parallel sets of p type AO's (px, py), as two atoms come together, there will be two equivalent sets of π MO's (πx and πy), lying in the xz and yz planes respectively (if z corresponds to the interatomic axis).

The following generic energy level diagram applies to all homonuclear diatomic molecules (with s and p valence AO's)

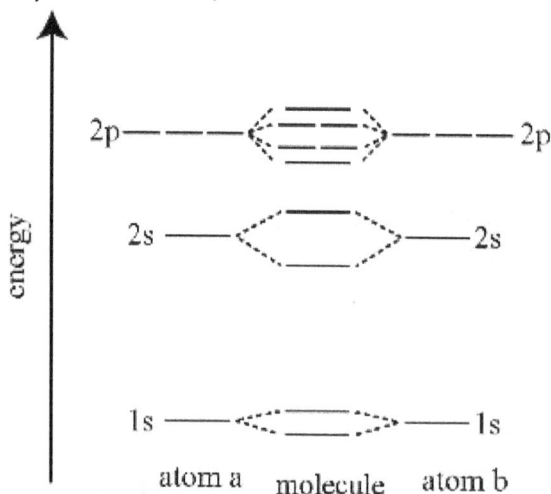

Next, to determine the ground state electronic configuration of the molecule, assign the electrons to the available molecular orbitals, as dictated by

- The Aufbau Rule (fill MO's in order of increasing energy)
- Pauli Exclusion Principle (a maximum of two electrons per MO with opposite spins)
- Hund's Rule (Maximize total spin when filling degenerate MO's) As an example, consider Li2

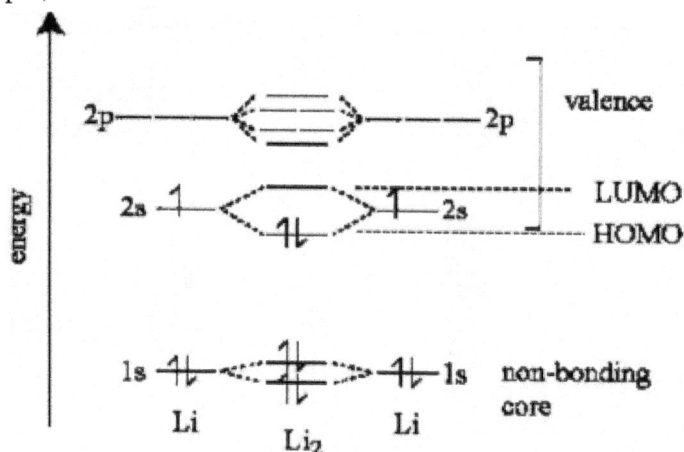

7.3 MOLECULAR ORBITAL DIAGRAMS

Molecular orbital (MO) diagrams are graphical representations that illustrate the formation of molecular orbitals through the combination of atomic orbitals in a molecule. These diagrams provide a visual depiction of the energy levels and electron occupancy of molecular orbitals, helping to explain the bonding, stability, and electronic properties of molecules. In a typical MO diagram, the atomic orbitals of constituent atoms are represented along the horizontal axis, with the energy levels increasing from left to right. The vertical axis denotes the energy levels of the resulting molecular orbitals, with lower energy levels located closer to the bottom of the diagram. The molecular orbitals are labelled according to their symmetry and bonding characteristics, such as σ (sigma), π (pi), and δ (delta) orbitals. The combination of atomic orbitals gives rise to bonding and antibonding molecular orbitals. Bonding molecular orbitals are formed through constructive interference, resulting in increased electron density between the nuclei of the bonded atoms, thus stabilizing the molecule. In contrast, antibonding molecular orbitals arise from destructive interference, leading to regions of reduced electron density between the nuclei and destabilizing the molecule. Molecular orbital diagrams typically depict the filling of electrons into the molecular orbitals according to the Aufbau principle, Hund's rule, and the Pauli exclusion principle. Electrons are filled into the molecular orbitals in order of increasing energy, with each orbital accommodating a maximum of two electrons with opposite spins. The relative energies of the molecular orbitals depend on factors such as the types of atomic orbitals involved, the overlap between the atomic orbitals, and the electronegativity of the atoms. For example, in diatomic molecules like hydrogen (H_2) or nitrogen (N_2), the σ bonding orbital is typically lower in energy than the π bonding orbital due to greater overlap of the atomic orbitals along the internuclear axis. Molecular orbital diagrams provide valuable insights into the electronic structure and bonding patterns of molecules, enabling researchers to predict and interpret their chemical behaviour, spectroscopic properties, and reactivity. These diagrams are essential tools in the fields of chemistry, physics, and materials science, facilitating the understanding and design of molecules with tailored properties for various applications.

7.3.1 Three simple kinds of molecular orbitals

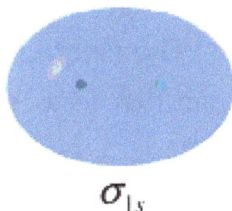

$$\sigma_{1s}$$

1. Sigma (bonding) orbitals.

Electrons delocalized around the two two nuclei. These may be represented as shared electrons, e.g. H:H or Li:Li

2. Non-bonding orbitals

Orbitals that are essentially unchanged from atomic orbitals, and remain *localized* on a single atom (unshared). These may be represented as a pair of electrons on one atom.

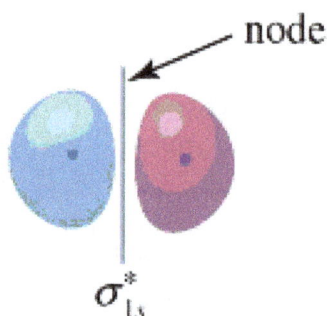

$$\sigma^{*}_{1s}$$

3. Sigma star (anti-bonding) orbitals

Orbitals with a node or nodes perpendicular to the axis between two nuclei. If occupied, these make a negative bonding contribution, i.e. cancel the contributions of occupied bonding orbitals.

7.3.2 Bond Order

Simple models of bonding include the concepts of *single, double,* and *triple bonds.*

Molecular orbital theory provides us with a natural and general definition of bond order that includes all of these and also intermediate bonds as follows:

Bond Order = ½ (No. of bonding electrons - No. of anti-bonding electrons)

E.g. H2 bond order = 1 (2 electrons in a σ MO)

Li2 bond order = 1 (2 electrons in a σ MO and 4 electrons in non-bonding core orbitals) H2+ bond order = 0.5 (1 electron in a σ MO)

H2- and He2+ bond order = 0.5 (2 electrons in a σ MO and 1 electron in a σ*MO)

He2 bond order = 0 (2 electrons in a σ MO and 2 electrons in a σ*MO)

7.3.3 Homonuclear diatomics: The electronic structure of N_2

Using an MO energy level diagram that focuses just on the valence electrons, allocate the 10 *valence* electrons of N2

Bond order = ½(8 - 2) = 3

There is an excess of 6 bonding electrons, corresponding to a triple bond: N≡N

The 10 valence electrons of N2 occupy bonding σ and π
MO's and an antibonding σ * MO.

The HOMO is actually a σ MO (lying slightly higher in energy than the σ MO's.)

The π* orbitals are empty - they are the (degenerate pair of) LUMO's.

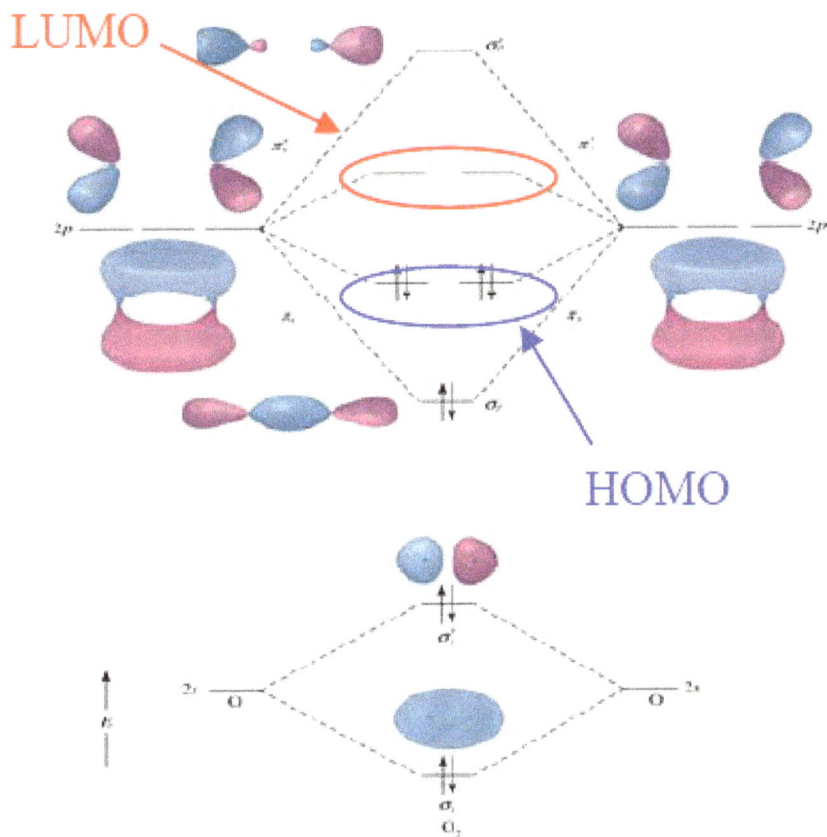

Homonuclear diatomics: The electronic structure of O2

O2 follows on from N2, the extra two electrons are placed in the degenerate pair of
π* MO's (as required by Hund's Rule).

Bond order = ½(8 - 4) = 2. This implies a double bond: O=O

MO theory also predicts that the O2 molecule would be paramagnetic, due to
the non-zero net electron spin, i.e. non-zero magnetic moment.

Oxygen is indeed paramagnetic!

In the next couple of examples, I have used the full MO diagram that includes the
core electrons as well as valence electrons. In all cases the same bond order is
obtained irrespecitce of whether the core electrons are included or not.

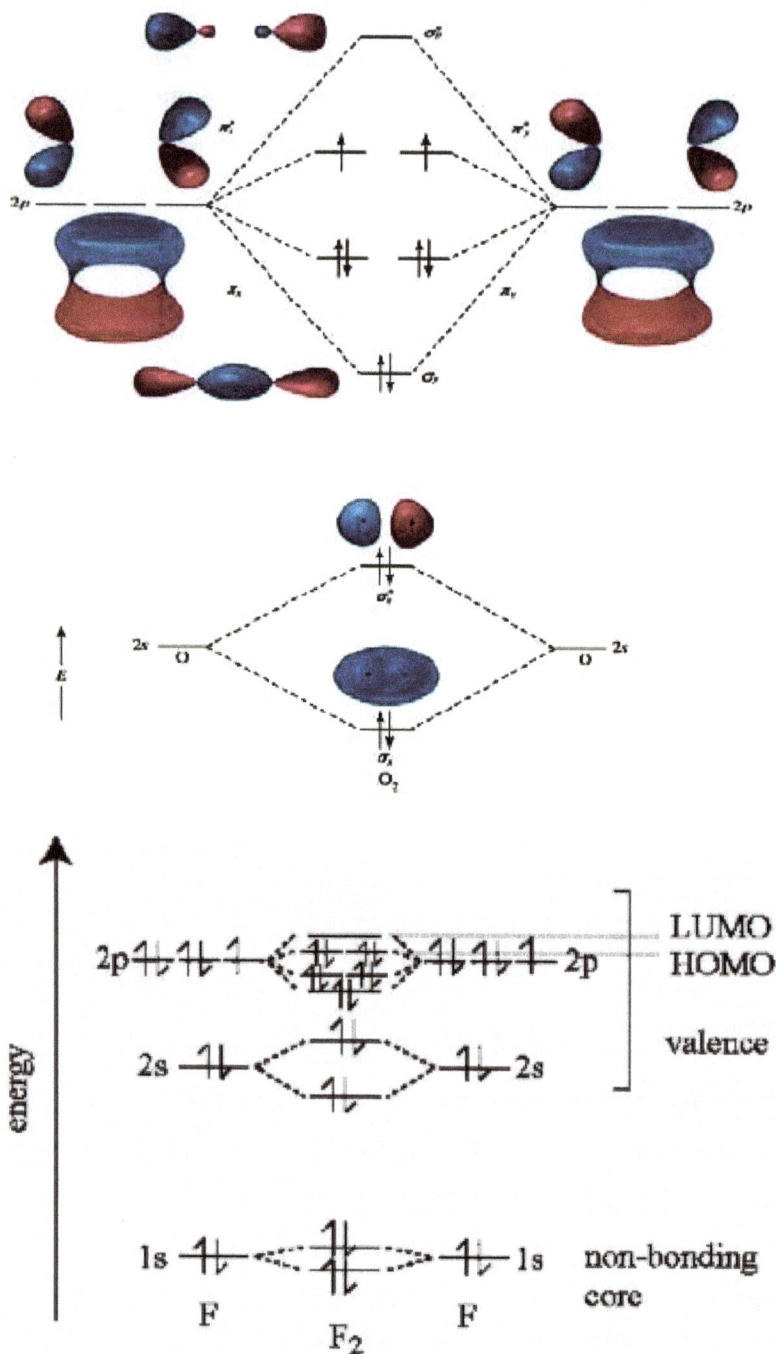

Homonuclear diatomics: The electronic structure of F2

In F2, the 18 electrons fill up all the MO's, up to and including the π * MO's.

Bond order = $\frac{1}{2}(8 - 6) = 1$. This implies a single bond: F-F

To obtain Ne2 the extra two electrons are placed in the $\pi *$ MO.

This results in a bond order of zero, i.e. no bond and no Ne2 molecule!

Heteronuclear diatomics: The electronic structure of NO

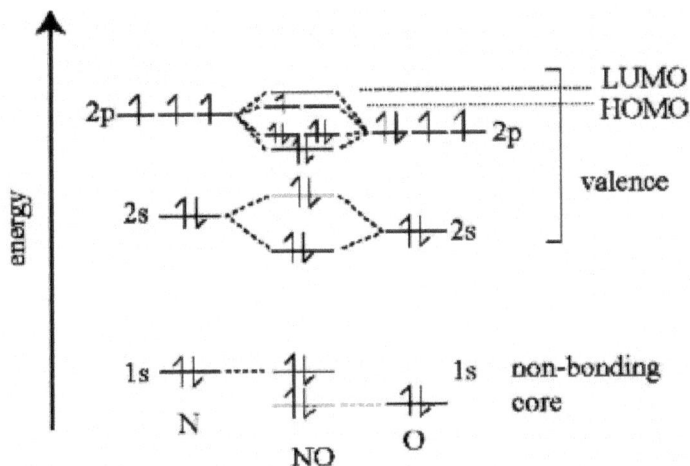

The energies of the AO's of N and O are very similar - those on O are slightly lower. The MO's of NO therefore can be constructed the same way as for N2 or O2.

Bond order = ½(8 - 3) = 2½. Strength of bond is between double and triple bonds. Molecule has an unpaired spin - therefore it is paramagnetic.

Heteronuclear diatomic: The electronic structure of HF

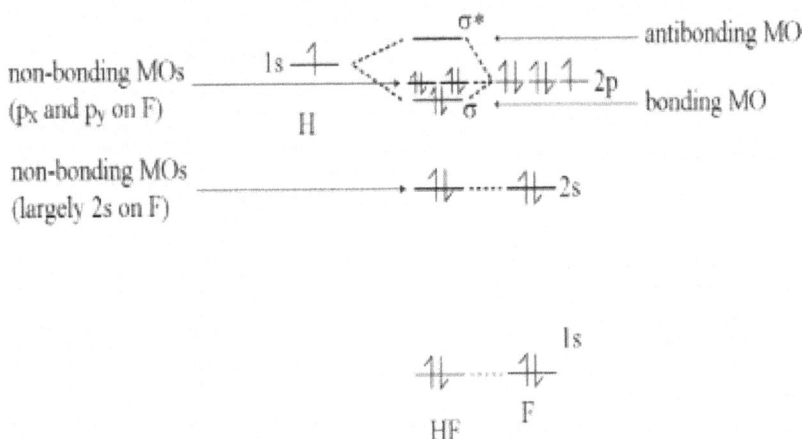

In hydrides, such as HF, the MO's need to be constructed from a single 1s AO of H and the 1s,2s,2p AO's of F. The 1s AO of H is closest in energy to the 2p AO's of F, but can only interact with the 2pz AO of F (because of symmetry). As a result, all doubly occupied AO's of F remain largely unchanged, as non-bonding orbitals.

MO's from interaction of s and p orbitals

$s \pm p_z \longrightarrow \sigma, \sigma^*$

(s and p_z AO's both have σ symmetry)

$s \pm p_x \longrightarrow 0$

(Zero overlap because s and p_x AO's
have σ and π symmetries respectively)

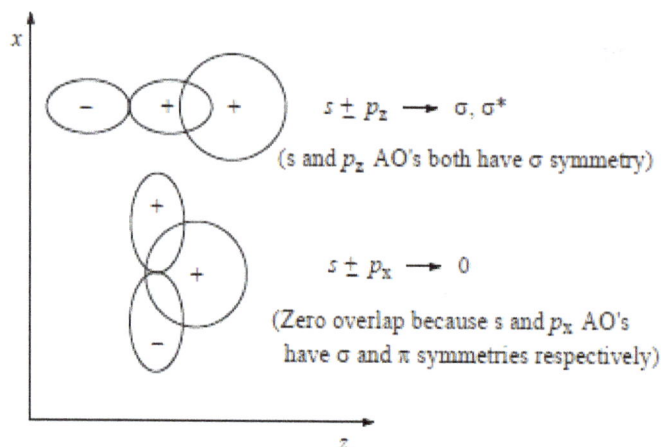

When forming MO's the parent AO's must have compatible symmetry!

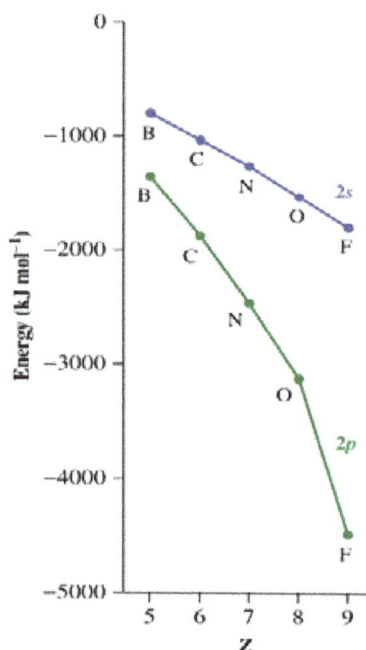

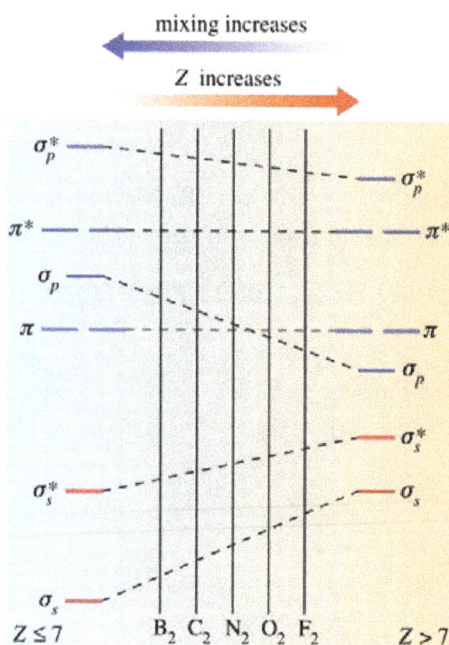

Supplementary information; you don't need to learn this, I will always provide you with the appropriate energy level diagram.

The relative energies of the atomic orbitals changes depending on atomic number. When the energy of the s and p atomic orbitals are close, the orbitals 'mix'.

The consequence of this is the ordering of the sigma molecular orbitals that arise from them changes relative to the pi orbitals which are derived entirely from the p orbitals.

Electron Densities in H2, F2, and HF

The square of a wavefunction (corresponding to an occupied orbital) tells us the charge density distribution of the electron(s) in the orbital. If we add up the charge densities from all the occupied molecular orbitals, we obtain the overall charge density distribution in the molecule.

This shows the surface for H_2 within which the probability of finding an electron is 95%. It is simply the square of the occupied σ MO.

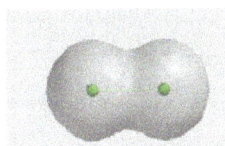

In F_2 the 95% surface includes all the occupied MO's. The *general effect* is seen by adding them together.

In HF the 95% surface looks like a simple sigma bond, but most of the electrons accumulate around the F atom.

The overall distribution of electron density in heteronuclear diatomic molecules is uneven due to the difference in nuclear charges and the different degree of attraction exerted on the electrons.

In NO the distribution of charge slightly favours O. In HF it strongly favours F. Similarly in HCl it favours Cl

Types of Orbitals and Bonds in Diatomics

We now know of five kinds of molecular orbitals formed by valence electrons.

⦿ σ (bonding) orbitals. Electrons in these bonds lower the energy of the molecule (relative to its atomic orbitals). These are shared between two nuclei and delocalised along the axis between two nuclei.

⦿ σ* (antibonding) orbitals. Electrons in these bonds raise the energy of the molecule (oppose bonding). These orbitals have a node or nodes along the axis between two adjacent nuclei.

⦿ Non-bonding (nb) orbitals are localised on only one atom and do not affect bonding.

⦿ π (bonding) orbitals. Electrons in these orbitals lower the energy of the molecule, and are delocalised between two nuclei in two lobes on opposite sides of the intern clear axis.

⦿ π* (antibonding) orbitals. These orbitals have lobes on opposite sides of the intern clear axis, and a node between adjacent atoms.

Bonding in Solids

1. Network Covalent, eg Diamond

The structure of diamond is known to be a tetrahedral arrangement of carbon atoms organised in a three-dimensional, crystalline array. This can be measured by e.g. x-ray diffraction, and the internuclear distances are known very precisely.

In our simple bonding model, every carbon atom in diamond is bonded to four carbon neighbours by a simple σ bond. The electrons are not delocalised further.

This model is a typical description of many materials we refer to as *network solids*. They are effectively large molecules with neighbouring atoms connected by a covalent σ bond.

C and Si are two elements that form covalent network crystals. Compounds that form covalent network solids include SiO_2, SiC, BN, and Si_3N_4.

Network solids like diamond can be treated as one large molecule, which means that the entire material has a set of quantum states (allowed energies), and that only two electrons can be in each orbital (allowed energy).

We can see the general effect of increasing molecular size by calculating the allowed energies in fragments of a 3-dimensional diamond network of increasing size. The allowed states fall into two groups, bonding and antibonding, as we would expect. As the number of atoms in the network structure increases, so does the number of allowed states and the *density of states* (how close together in energy they are).

The ground state electronic configuration of network solids has all the σ energy levels filled, and all of the σ* energy levels empty.

The lowest energy (HOMO → LUMO) electronic transition is given by the *band gap*, the energy difference between the top of the (filled) band of allowed σ energies and the (empty) band of allowed σ * energies.

In network solids and insulators, this band-gap energy is very large.

These materials are colourless and transparent because the longest wavelength that can be absorbed is shorter than the shortest wavelength in the visible spectrum (approx. 400 nm)

That is, $E_{band-gap} > \dfrac{hc}{\lambda_{min}} = \dfrac{6.626 \times 10^{-34} \times 3.00 \times 10^{8}}{4.0 \times 10^{-7}}$

Eband-gap > 5.0 x 10-19J or 3.1eV

2. Metallic Bonding

Metals are also crystals in which the atoms are bonded to one another and can be treated as a single, large molecule. However in metals the bands of allowed energy levels are remarkably different from insulators.

If we take the same approach with, say sodium, as for diamond, we find that increasing the size of the fragment gives two bands of energy levels with <u>no band gap</u>. Energy levels in metals behave as a single, *partially-filled* band.

This means that there are many energy levels close together, and that the longest wavelength transition is much longer than 400nm, so the materials are <u>opaque</u>.

Natural Semiconductors are network solids with band gap energies that lie in the visible or UV range. They may thus be transparent (UV absorbing) or coloured (visible absorbing). Absorption of a photon promotes an electron from the lower, filled band into the unfilled upper band. Once in this band (the conduction band), the electron has enough thermal energy to *move* and hence to conduct electricity.

Promotion of an electron leaves a *vacancy* or hole in the lower (valence) band, so electrons there <u>also become mobile</u>, and have enough thermal energy to move between states within that band.

Conduction can be regarded as taking place through both electrons in the conduction band and holes in the valence band.

Electrons can be promoted into conduction band states by light, or by thermal excitation (heat).

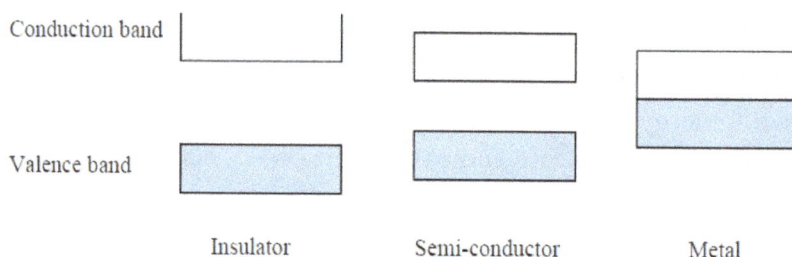

Insulator   Semi-conductor   Metal

In natural semiconductors with small band gaps, some electrons are thermally excited into the conduction band. The fraction of excited electrons increases with temperature, _and so does the conductivity_.

Materials that are insulators at low temperatures become increasingly good semi- conductors with increasing temperature.

Doped Semiconductors

Semiconductors can be synthesised by introducing foreign atoms into an insulator to modify its electronic structure. There are two types of doped semiconductors.

N-type semiconductors are prepared by introducing atoms with occupied quantum states _just below the bottom of the conduction band._

Some electrons from these _localised_ electronic states are thermally excited into the conduction band, where they become mobile and act as (negative) charge carriers.

Typical n-type semiconductors are prepared by substituting group 15 elements (P, As, Sb) into the crystal lattice of Si or Ge (group 14). Group 16 elements can act as double donors into these lattices.

P-type semiconductors are prepared by introducing atoms with vacant quantum states _just above the top of the valence band._

Some electrons from the filled valence band are thermally excited into these localised orbitals. This leaves vacancies or holes in the valence band that are mobile and act as (positive - "p-type") charge carriers.

Typical p-type semiconductors are prepared by substituting group 13 (B, Al, Ga) or group 2 or 12 (Be or Zn) elements into the crystal lattice of an insulator.

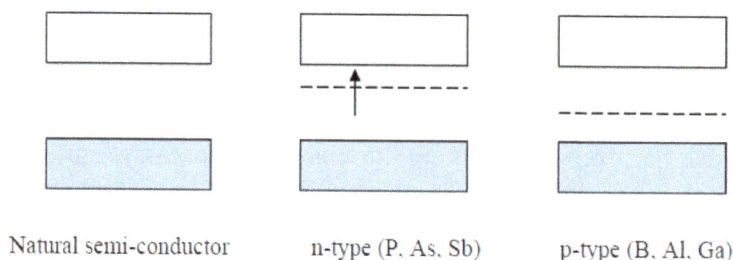

Natural semi-conductor   n-type (P. As, Sb)   p-type (B. Al, Ga)

Substitution into compound semiconductors - e.g. GaAs rather than Si or Ge - are a little more complex. For example, Group14 additives can act as donors or acceptors, depending on which element they substitute.

3. Ionic "Bonding"

Molecular Orbitals come about when the energy of delocalised valence electrons (bonding MO's) are lower than those localised on individual atoms.

In extreme cases where the allowed energies of electrons in two different atoms are very different, the lowest energy state of the two atoms together is not a bond but the transfer of one or more electrons from one atom to an atomic orbital of another.

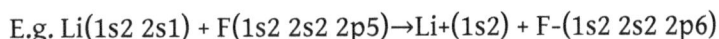

E.g. Li(1s2 2s1) + F(1s2 2s2 2p5)→Li+(1s2) + F-(1s2 2s2 2p6)

This kind of electron transfer leads to the formation of two ions.

In order for this to be favourable (even more favourable than delocalisation into a MO), the available atomic orbital of the acceptor atom must be much lower in energy than the highest filled atomic orbital of the donor atom. This usually means few outer shell electrons for the donor (big atom) and an almost filled outer shell for the acceptor (small atom with tightly bound electrons)

Electronegativity again proves to be a useful concept in dealing with ionic bonds. From the periodic table we can see that the least electronegative atoms are good electron donors (cation formers), and the most electronegative atoms are good electron acceptors (anion formers).

We can use the electronegativity difference between two atoms (ΔEN) to empirically define the *partial ionic character* of a bond as a fraction of the maximum possible difference, 4.0.

E.g. for HF, $\Delta EN = 4.0 - 2.1 = 1.9$

Partial Ionic Character = $1.9/4.0 = 0.495$

HCl: $(3.0 - 2.1)/4.0 = 0.23$

NO: $(3.5 - 3.0)/4.0 = 0.13$

LiF: $(4.0 - 1.0)/4.0 = 0.75$

MgCl2: $(3.0 - 1.2)/4.0 = 0.45$

Electronegativity differences >2 generally give ionic bonds, whereas $\Delta EN < 1$ are covalent (delocalised MO's). This gives a good guide to the *character* of a bond.

What we call an ionic bond is simply the long-range electrostatic attraction between cation(+) and anion(-), together with the short-range repulsion between electrons in adjacent ions. The equilibrium distance between cation and anion nearest-neighbours occurs when the <u>potential</u> energy is a minimum. That is, when the attractive and repulsive forces are exactly equal and opposite.

7.4 APPLICATION OF SIMPLE DIATOMIC MOLECULES

Simple diatomic molecules, such as hydrogen (H2), oxygen (O2), nitrogen (N2), and chlorine (Cl2), find extensive applications across diverse scientific and industrial domains. These molecules serve as fundamental building blocks for understanding chemical principles, materials synthesis, environmental monitoring, and technological advancements. In fields like chemistry and physics, diatomic molecules provide essential models for studying molecular structure, bonding, and spectroscopic properties, offering insights into the behavior of more complex molecular systems. One significant application of diatomic molecules lies in their role as model systems for elucidating basic chemical concepts. They serve as ideal platforms for teaching fundamental principles of chemical bonding, molecular symmetry, and quantum mechanics due to their simplicity and well-understood properties. Additionally, diatomic molecules play crucial roles in quantum chemistry studies, where they serve as benchmark systems for developing and testing computational methods to describe molecular behavior with high accuracy. Furthermore, diatomic molecules have practical applications in spectroscopy, where their characteristic absorption and emission spectra provide valuable information about molecular structure and dynamics. Techniques such as infrared spectroscopy and rotational spectroscopy rely on the unique spectral signatures of diatomic molecules to analyze molecular compositions, monitor chemical reactions, and identify substances in various samples. These spectroscopic methods have broad applications across chemistry, physics, environmental science, and biomedical research.

In industrial settings, diatomic molecules serve as key components in numerous manufacturing processes and technological applications. For instance, hydrogen and oxygen are essential reactants in the production of ammonia for fertilizer synthesis, while nitrogen gas finds use in inert atmospheres for preserving sensitive materials and preventing oxidation. Chlorine is widely employed in water treatment facilities for disinfection purposes and in the synthesis of numerous organic compounds essential for pharmaceuticals, plastics, and agricultural chemicals. Moreover, diatomic molecules play critical roles in atmospheric and environmental sciences. The abundance and reactivity of oxygen and nitrogen in the Earth's atmosphere influence climate patterns, atmospheric chemistry, and ecological processes. Understanding the behavior of diatomic molecules in the atmosphere is crucial for assessing air quality, monitoring pollutant levels, and studying the impact of human activities on environmental health. The application of principles to simple diatomic molecules involves understanding their structure, bonding, spectroscopy, and other properties. Let's delve into each aspect:

- **Molecular Structure:** Diatomic molecules consist of two atoms of the same element bonded together. Examples include hydrogen (H2), oxygen (O2), nitrogen (N2), and chlorine (Cl2). The bond between the two atoms is typically covalent, where electrons are shared between the atoms. The molecular structure is typically linear, with a bond angle of 180 degrees.

- **Chemical Bonding:** The bonding in diatomic molecules is primarily covalent, where the atoms share electrons to complete their valence shells. This sharing creates a bond between the atoms, stabilizing the molecule. The strength of the bond depends on factors such as the types of atoms involved and the overlap of their atomic orbitals.

- **Spectroscopy:** Spectroscopy techniques, such as infrared spectroscopy and rotational spectroscopy, are crucial for studying diatomic molecules. Infrared spectroscopy measures the vibrational modes of molecules by detecting the absorption of infrared radiation. Rotational spectroscopy provides information about the rotational energy levels of molecules.

- **Quantum Mechanics:** Quantum mechanics plays a significant role in understanding the behavior of diatomic molecules. Concepts such as molecular orbitals, wave functions, and the Schrödinger equation are used to describe the electronic structure and properties of these molecules.

- **Molecular Energy Levels:** Diatomic molecules have discrete energy levels associated with their electronic, vibrational, and rotational states. The quantization of energy levels leads to characteristic spectra observed in spectroscopic experiments.

- **Molecular Symmetry:** Symmetry considerations are essential for understanding the selection rules governing transitions between different molecular states. Symmetry operations, such as rotation and reflection, help classify molecular orbitals and predict spectroscopic outcomes.

- **Chemical Reactivity:** Diatomic molecules participate in various chemical reactions, including oxidation, reduction, and formation of new bonds. The reactivity of these molecules depends on factors such as bond strength, electronegativity, and molecular orbital interactions.

Understanding these aspects allows scientists to predict and interpret the behavior of diatomic molecules in various experimental conditions, providing insights into fundamental chemical processes and applications in fields such as materials science, environmental science, and medicine.

08 | VALENCE BOND THEORY

8.1 INTRODUCTION: AN OVERVIEW

We have already said that Werner was the first to explain the nature of bonding in complex compounds. However, with the advancement of theories of valence, modern theories have been proposed to explain the nature of metal-ligand bonding in complexes. These theories can also explain the colour, geometry and magnetic properties of the complex compounds. These mode theories are:

- ⊙ **Valence Bond Theory:** VBT (due to L. Pauling and JL Slater, 1935)
- ⊙ **Crystal Field Theory:** CFT (due to H. Bethe. 1929 and Van Vleck, 1932) (iii) Ligand Field Theory LFT or Molecular Orbital Theory, MOT (due to J. Van Vleck, 1935). Here we shall discuss only valence bond theory.

8.2 VALENCE BOND THEORY (VBT)

This theory is mainly due to Pauling. It deals with the electronic structure of the central metal ion in its ground state, kind of bonding, geometry and magnetic properties of the complexes.

Assumptions of valence bond theory :

1. The central metal atom or ion makes available a number empty s, p and d atomic orbitals equal to its coordination number. These vacant orbitals hybridized together to form hybrid orbitals which are the same in number as the atomic orbitals hybridized together. These hybrid orbitals are vacant, equivalent in energy and have definite geometry: Important types of hybridisation occuring in the first row transition metal (3d) complexes and the geometry of the complex are given in Table 1.

1. The ligands have at least one σ-orbital containing a lone pair of electrons. Vacant hybrid orbitals of the metal atom or ion overlap with the filled (containing lone pair of electrons) σ-orbitals of the ligands to form ligand→ metal σ-bond. This bond is known as coordinate bond is a special type of covalent bond and shows the characteristics of both the overlapping orbitals. However, it also possesses a considerable amount of polarity because of the mode of its formation.

Table 1. Important types of hybridisation found in the first row transition metal complexes and the geometry of the complexes

Coordination number of the central metal atom/ion	Type of hybridisation undergone by the central metal atom/ion	Geometry of the complex	Examples of complexes
2	$sp(4s, 4p_x)$	Linear or diagonal	$[CuCl_2]^-$, $[Cu(NH_3)_2]^+$ etc.
3	$sp^2(4s, 4p_x, 4p_y)$	Trigonal planar or equilateral triangular	$[Cu^+Cl(tu)_2]^0$ (distorted trigonal planar) etc.
4	$dsp^2(3d_{x^2-y^2}, 4s, 4p_x, 4p_y)$	Square planar	$[Ni(CN)_4]^{2-}$, $[PdCl_4]^{2-}$
4	$sp^2d(4s, 4p_x, 4p_y, 4d_{x^2-y^2})$	Square planar	$[Cu(NH_3)_4]^{2+}$ $[Pt(NH_3)_4]^{2+}$ etc.
4	$sp^3(4s, 4p_x, 4p_y, 4p_z)$	Tetrahedral	$[NiCl_4]^{2-}$, $[Cu(CN)_4]^{3-}$, $Ni(CO)_4$ etc.
5	$dsp^3(3d_{z^2}, 4s, 4p_x, 4p_y, 4p_z)$	Trigonal bipyramidal	$Fe(CO)_5$, $[CuCl_5]^{3-}$, $[Ni^{2+}(triars) Br_2]^0$
5	$dsp^3(3d_{x^2-y^2}, 4s, 4p_x, 4p_y, 4p_z)$	Square pyramidal	$[Co^{2+}(triars) I_2]^0$, $[Ni(CN)_5]^{3-}$ etc.
6	$d^2sp^3(3d_{x^2-y^2}, 3d_{z^2}, 4s, 4p_x, 4p_y, 4p_z)$	Inner-orbital octahedral	$[Ti(H_2O)_6]^{3+}$, $[Fe(CN)_6]^{3-}$ etc.
6	$sp^3d^2(4s, 4p_x, 4p_y, 4p_z, 4d_{x^2-y^2}, 4d_{z^2})$	Outer-orbital octahedral	$[Fe^+(NO^+)(H_2O)_5]^{2+}$, $[CoF_6]^{3-}$ etc.

2. The non-bonding electrons of the metal atom or ion are rearranged in the metal orbitals (viz. pure d, s or p orbitals as the case may be) which do not participate in forming the hybrid orbitals. The rearrangement of non-bonding electrons takes place according to Hund's rule.

8.3 GEOMETRY OF 6-COORDINATED COMPLEX IONS

In all the complex ions the coordination number of the central metal atom or ion is six the complex ions have octahedral geometry.

This octahedral geometry arises due to d2sp3 or sp3d2 hybridisation of the central metal atom or ion. What type of hybridisation will occur depends on the number of unpaired or paired electrons present in the complex ion. d2sp3 or sp3d2 hybridisation is also called octahedral hybridisation.

Octahedral complexes in which the central atom is d2sp3 hybridised are called inner-orbital octahedral complexes while the octahedral complexes in which the central atom is sp3d2 hybridised are called outer-orbital octahedral complexes.

8.3.1 d2sp3 Hybridisation in Inner Orbital Octahedral Complexes:

This type of hybridisation takes place in those octahedral complexes which contain strong ligands. On the basis of the orientation of the lobes of d-orbitals in space, these orbitals have been classified into two sets viz. t2g (dxy, dyz and dxz) and eg (dz2and dx2-y2) orbitals. In the formation of six d2sp3 hybrid orbitals, two (n-1) d-orbitals of eg set, one ns and three np (npx, npy, and npz) orbitals combine together and form six d2sp3 hybrid orbitals.

Thus we see that the two d-orbitals used in d2sp3 hybridisation are from penultimate shells [ie. (n-1)th shell] while s and three p-orbitals are from ultimate shell (ie. nth shell). This discussion shows that in case of octahedral complex ions of 3d transition series elements, two d-orbitals used in d2sp3 hybridisation are 3dz2 and 3dx2-y2 orbitals while s- and p-orbitals are 4s and 4p orbitals. Thus d2sp3 hybridization taking place in such complexes can be represented as: dx2-y2. dz2.4s.4px.4py.4pz(d2sp3).

Since two d-orbitals used in d2sp3 hybridisation belong to the inner shell [i.e. (n-1)th shell], the octahedral complex compounds resulting from d2sp3 hybridisation are called **inner orbital octahedral complexes.**

8.3.2 Ferrocyanide ion OR hexacyanoferrate (II) ion [Fe(CN)$_6$]4- :

In this ion, since the coordination number of Fe is six, the given complex ion has octahedral geometry. In this ion, Fe is present as Fe2+ ion whose valence-shell configuration is 3d6 4s0 4p0 or , t2g eg 4s 4p which shows that Fe2+ ion has 4 unpaired electrons. Magnetic studies have, however, shown that the given complex ion is diamagnetic and hence it has no unpaired electrons (n = 0). Hence in order to get all the electrons in the paired state, two electrons of eg orbitals are sent to t2g orbitals so that n becomes equal to zero.

Since CN- ions (ligands) are strong ligands, they are capable of forcing the two electrons of eg orbitals to occupy t2g orbitals and thus make all the electrons paired. Now for the formation of [Fe(CN)6]4- ion, two 3d orbitals of eg set, 4s orbital (one orbital) and three 4p orbitals (all these six orbitals are vacant orbitals) undergo d2sp3 hybridisation (see Figure 1). It is due to d2sp3 hybridisation that [Fe(CN)6]4- ion is an inner orbital octahedral complex ion. The electron pair donated by CN- ion (ligand) is accommodated in each of the six d2sp3 hybrid orbitals as shown in Figure.

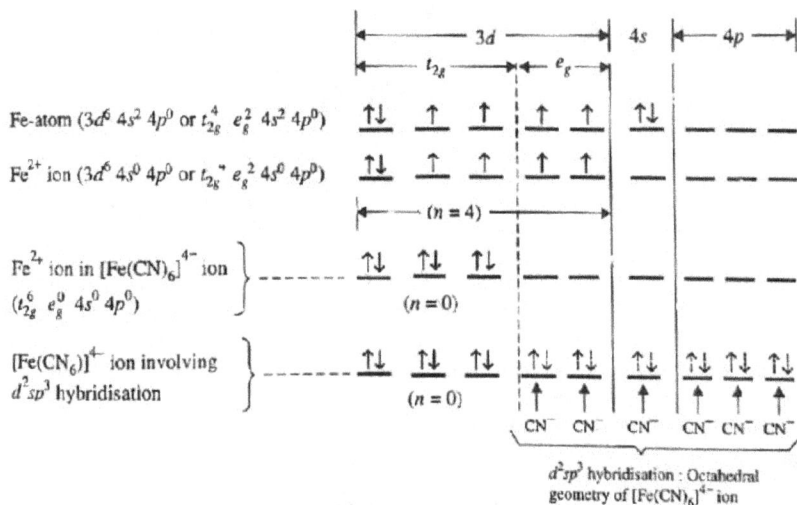

Fig. 1: Formation of [Fe(CN)$_6$]$^{4-}$ ion by d^{2}sp^3 hybridisation. ↑↓ indicates electron pair donated by each CN ion (ligand) (Inner-orbital octahedral complex ion).

8.3.3 sp3d2 Hybridisation in Outer Orbital Octahedral Complexes

This type of hybridisation takes place in those octahedral complex ions which contain weak ligands. Weak ligands are those which cannot force the electrons of dz2 and dx2-y2 orbitals of the inner shell to occupy dxy, dyz and dxz orbitals of the same shell. Thus in this hybridisation. (n-1)dz2 and (n-1)dx2-y2 orbitals are not available for hybridisation. In place of these orbitals, we use ndz2 and ndx2-y2 orbitals (These d-orbitals belong to the outer shell) and hence sp3d2 hybridisation can be represented as ns, npx, npy, npz, nd2, ndx2-y2 .

This hybridisation shows that all the six orbitals involved in hybridisation belong to the higher energy level (outer shell). This discussion shows that in the case of octahedral ions of 3d transition series, d-orbitals used in hybridisation are 4dz2 and 4dx2-y2 orbitals. Since two d-orbitals are from the outer shell (i.e., nth shell), the octahedral complexes resulting from sp3d2 hybridisation are called **outer orbital octahedral complexes**. Since these complexes have a comparatively greater number of unpaired electrons than the inner orbital octahedral complexes, these are also called **high spin or spin free octahedral** complexes. Now let us discuss the structure of some octahedral complex ions of 3d transition series elements which are formed by sp3d2 hybridisation.

8.3.4 Hexafluroferrate (III) ion [FeF$_6$]3- :

In this ion, the coordination number of Fe is six and hence the given complex ion has octahedral geometry. Here iron is present as Fe3+ whose valence shell electronic configuration is 3d5 4s0 4p0 or t2g2eg .

Each of the five electrons is unpaired and hence n = 5. Magnetic properties of the given ion have also shown that the ion has five unpaired electrons and hence is paramagnetic corresponding to the presence of five unpaired electrons. Thus two electrons residing in orbitals cannot be forced to occupy t2g orbitals as we have done in case of [Fe(CN)6]3- ion, otherwise the number of electrons would become equal to one.

Thus we find that in case of the given ion, the two d-orbitals used in hybridisation are 4dz2 and 4dx2-y2, (and not 3dz2 and 3dx2-y2 as in case of [Fe(CN)6]3- ion) and s and p orbitals are 4s and 4p. Thus the given ion results from (4s) (4p3) (4dz2) (4dx2-y2) hybridisation as shown in Figure 1. This discussion shows that in the formation of [FeF6]3- in the original valence-shell configuration of Fe3+ ion is not disturbed.

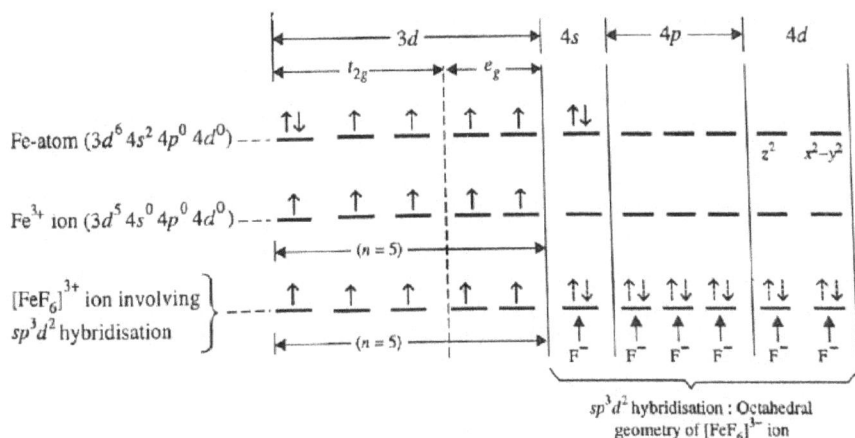

Fig. 1: Formation of $[FeF_6]^{3-}$ ion by $sp^3 d^2$ hybridisation. (Outer-orbital Complex ion).

Differences between inner orbital and outer orbital octahedral complexes

Inner orbital octahedral complexes	Outer orbital octahedral complexes
1. These are formed by d2sp3 hybridisation i.e.in the formation of six d2sp3 hybrid orbitals, two (n-1)d orbitals of eg set, one ns and three np orbitals are used. (n-1)d orbitals belong to the inner (penultimate) shell while ns and	1. These are formed by sp3d2 hybridisation i.e.in the formation of six sp3d2 hybrid orbitals, two nd orbitals of eg set, one ns and three np orbitals are used. Thus all the orbitals belong to the outer (ultimate) shell.
2. These complexes have comparatively lesser number of unpaired electrons and hence are also called low spin or spin paired complexes.	2. These complexes have a comparatively greater number of unpaired electrons and hence are also called high spin or spin free complexes.
3. These are given by strong ligands.	3. These are given by weak ligands.

8.4 GEOMETRY OF 4-COORDINATE COMPLEX IONS

Examples of 4-coordinated complex ions formed by some transition metals. In these complex ions the coordination number of the central metal atom or ion is four. Such complex ions may have either square planar or tetrahedral geometry, depending on whether the central atom or ion is dsp2 or sp3 hybridised. What type of hybridisation (ie.. whether dsp2 or sp3) the central metal atom or ion of a 4-coordinated complex ion undergoes depends on the number of unpaired or paired electrons present in the complexion.

8.4.1 Square Planar geometry (dsp²)

[Ni(CN) 4] 2- Tetracyanonickelate(II)ion

When [NI(CN)4]²⁻ ion is square planar geometry, Ni²⁺ ion should be dsp 2 hybridised. In this hybridisation, due to the energy made available by the approach of four CN - ions (ligands), the two unpaired 3d-electrons are paired up, thereby, making one of the 3d orbitals empty. This empty 3d orbital (which is 3d x2-y2 orbital) is used in dsp 2 hybridisation. This hybridisation makes all the electrons paired (n = 0) and hence is diamagnetic, as shown below Figure.

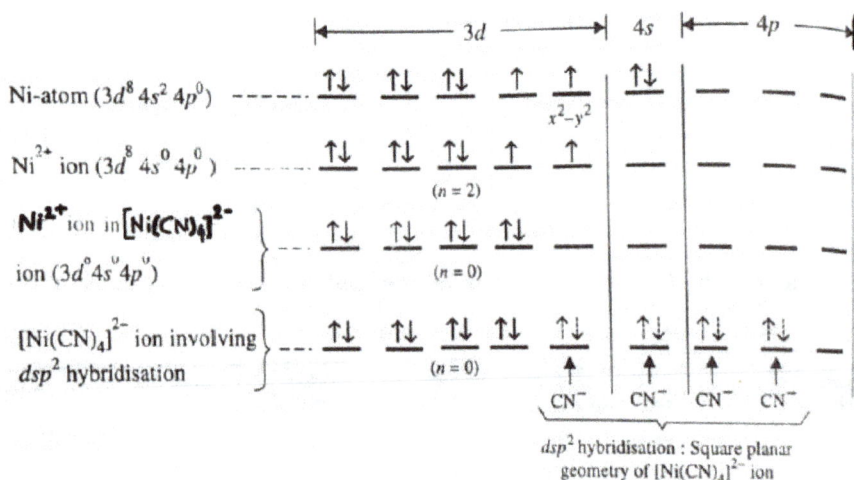

dsp² hybridisation : Square planar geometry of [Ni(CN)₄]²⁻ ion

Fig. 2: Formation [Ni(CN)₄]²⁻ ion by *dsp²* hybridisation (Square planar complex ion with *n = 0*)

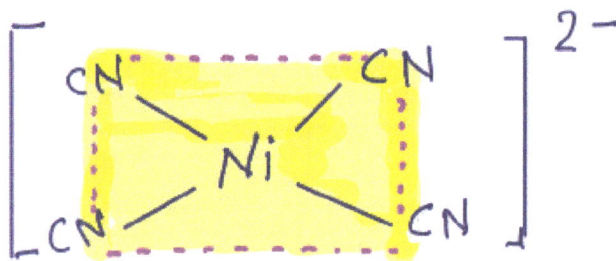

8.4.2 Tetrahedral geometry (sp ^{3})

A. [Ni(CO) 4] molecule (Tetracarbonylnickel) :

In this complex compound Ni is in zero oxidation state and has its valence-shell configuration as 3d8 4s2 . This compound has tetrahedral geometry which arises due to sp hybridisation of Ni atom. The magnetic studies of [Ni(CO)$_4$] have indicated that this molecule is diamagnetic (n = 0), showing that the two 4s electrons are forced to pair up with 3d orbitals. This results in sp^3 hybridisation and the [Ni(CO)4] molecule has a tetrahedral structure in the figure 3.

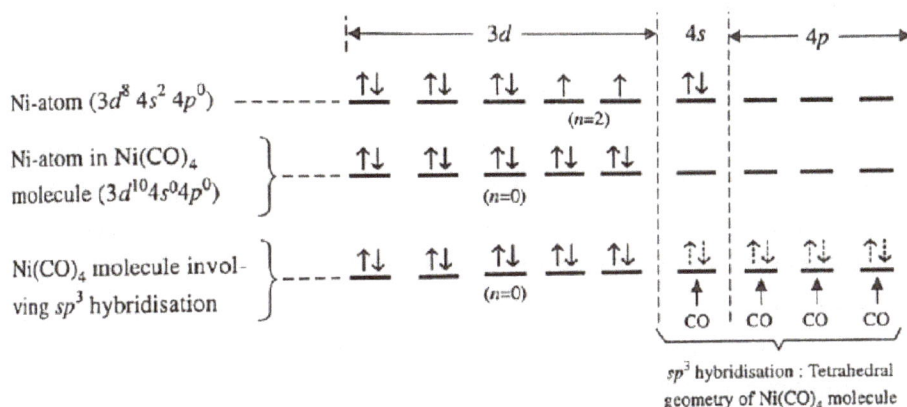

Fig. sp^3 hybridisation of Ni-atom in Ni(CO)4 molecule which has tetrahedral shape.

Fig. 3: sp^3 hybridisation of Ni-atom in Ni (CO)$_4$ molecule which has tetrahedral shape.

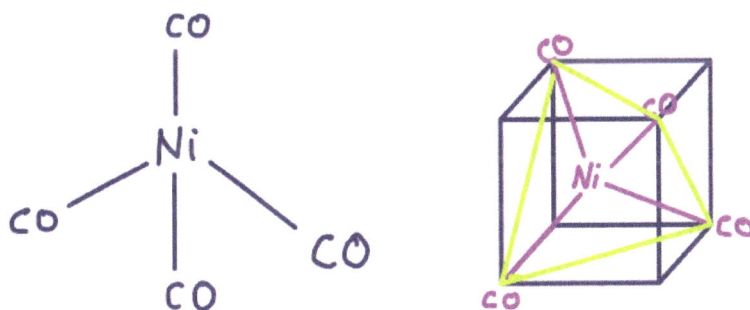

B. [NICl 4] 2- ion (Tetrachloronickelate(II)) :

This complex ion has Ni 2+ ion whose valence-shell configuration is 3d 8 4s 0 . Magnetic measurements reveal that the given ion is paramagnetic and has two unpaired electrons (n = 2). This is possible only when this ion is formed by sp 3 hybridisation and has tetrahedral geometry in the figure.

09 CRYSTAL LATTICE STRUCTURE AND BONDING

9.1 INTRODUCTION: AN OVERVIEW

The crystal lattice structure and bonding represent fundamental concepts in the realm of materials science and solid-state physics, governing the properties and behavior of a vast array of substances. At its core, the crystal lattice embodies the ordered arrangement of atoms, ions, or molecules within a crystalline solid, forming a three-dimensional network characterized by repetitive patterns and symmetries. This arrangement not only defines the macroscopic shape and structure of crystals but also profoundly influences their mechanical, thermal, optical, and electronic properties. Central to the understanding of crystal lattice structures is the concept of bonding, which delineates the forces that hold the constituent particles together within the lattice. These bonds arise from the interactions between neighboring atoms, ions, or molecules, and their nature can vary widely depending on the types of elements involved and the specific conditions under which the crystal forms. Common types of bonding mechanisms include ionic, covalent, metallic, and van der Waals interactions, each imparting distinct characteristics to the resulting crystal lattice. Through the exploration of crystal lattice structures and bonding mechanisms, scientists and engineers gain invaluable insights into the properties and potential applications of crystalline materials across diverse fields, including semiconductors, ceramics, metals, and polymers. By elucidating the intricate interplay between atomic arrangement and bonding forces, researchers can manipulate and tailor the properties of crystalline solids to meet the demands of modern technology, paving the way for innovative advancements in materials design, manufacturing, and performance.

9.2 CRYSTAL LATTICE STRUCTURES

Crystalline solids have regular ordered arrays of components held together by uniform intermolecular forces, whereas the components of amorphous solids are not arranged in regular arrays. With few exceptions, the particles that compose a solid material, whether ionic, molecular, covalent, or metallic, are held in place by strong attractive forces between them. When we discuss solids, therefore, we consider the positions of the atoms, molecules, or ions, which are essentially fixed in space, rather than their motions (which are more important in liquids and gases). The constituents of a solid can be arranged in two general ways: they can form a regular repeating three-dimensional structure called a crystal lattice, thus producing a crystalline solid, or they can aggregate with no particular order, in which case they form an amorphous solid (from the Greek ámorphos, meaning "shapeless").

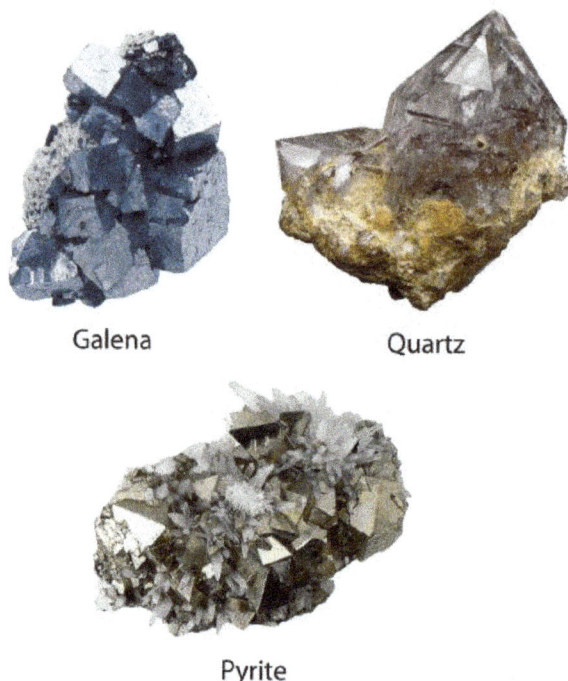

Galena

Quartz

Pyrite

Figure 1: (left) Crystalline faces. The faces of crystals can intersect at right angles, as in galena (PbS) and pyrite (FeS2), or at other angles, as in quartz. (Right) Cleavage surfaces of an amorphous solid. Obsidian, a volcanic glass with the same chemical composition as granite (typically KAlSi3O8), tends to have curved, irregular surfaces when cleaved.

Crystalline solids, or crystals, have distinctive internal structures that in turn lead to distinctive flat surfaces, or faces. The faces intersect at angles that are characteristic of the substance. When exposed to x-rays, each structure also produces a distinctive pattern that can be used to identify the material. The characteristic angles do not depend on the size of the crystal; they reflect the regular repeating arrangement of the component atoms, molecules, or ions in space. When an ionic crystal is cleaved (Figure 2, for example, repulsive interactions cause it to break along fixed planes to produce new faces that intersect at the same angles as those in the original crystal. In a covalent solid such as a cut diamond, the angles at which the faces meet are also not arbitrary but are determined by the arrangement of the carbon atoms in the crystal.

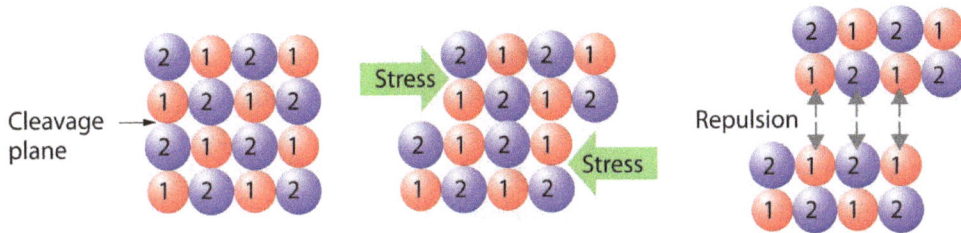

Figure .2: Cleaving a Crystal of an Ionic Compound along a Plane of Ions. Deformation of the ionic crystal causes one plane of atoms to slide along another. The resulting repulsive interactions between ions with like charges cause the layers to separate.

Crystals tend to have relatively sharp, well-defined melting points because all the component atoms, molecules, or ions are the same distance from the same number and type of neighbours; that is, the regularity of the crystalline lattice creates local environments that are the same. Thus, the intermolecular forces holding the solid together are uniform, and the same amount of thermal energy is needed to break every interaction simultaneously.

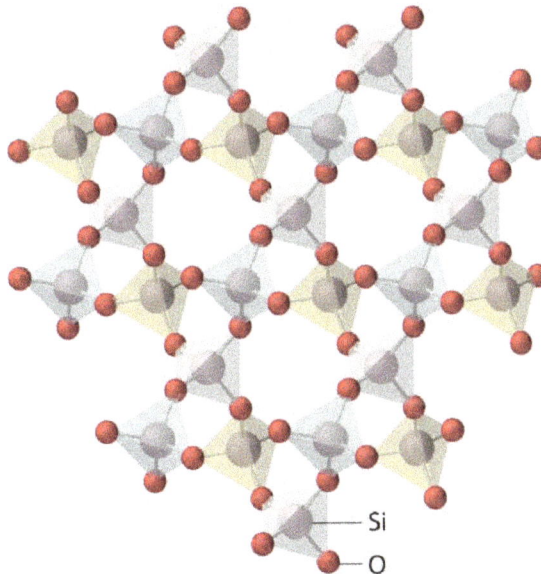

Figure 3: The lattice of crystalline quartz (SiO_2). The atoms form a regular arrangement in a structure that consists of linked tetrahedra.

Amorphous solids have two characteristic properties. When cleaved or broken, they produce fragments with irregular, often curved surfaces; and they have poorly defined patterns when exposed to x-rays because their components are not arranged in a regular array. An amorphous, translucent solid is called a glass. Almost any substance can solidify in amorphous form if the liquid phase is cooled rapidly enough. Some solids, however, are intrinsically amorphous, because either their components cannot fit together well enough to form a stable crystalline lattice or they contain impurities that disrupt the lattice. For example, although the chemical composition and the basic structural units of a quartz crystal and quartz glass are the same—both are SiO_2 and both consist of linked SiO_4 tetrahedra—the arrangements of the atoms in space are not. Crystalline quartz contains a highly ordered arrangement of silicon and oxygen atoms, but in quartz glass the atoms are arranged almost randomly. When molten SiO_2 is cooled rapidly (4 K/min), it forms quartz glass, whereas the large, perfect quartz crystals sold in mineral shops have had cooling times of thousands of years. In contrast, aluminum crystallizes much more rapidly. Amorphous aluminum forms only when the liquid is cooled at the extraordinary rate of 4×10^{13} K/s, which prevents the atoms from arranging themselves into a regular array.

In an amorphous solid, the local environment, including both the distances to neighboring units and the numbers of neighbors, varies throughout the material. Different amounts of thermal energy are needed to overcome these different interactions. Consequently, amorphous solids tend to soften slowly over a wide temperature range rather than having a well-defined melting point like a crystalline solid. If an amorphous solid is maintained at a temperature just below its melting point for long periods of time, the component molecules, atoms, or ions can gradually rearrange into a more highly ordered crystalline form.

Crystals have sharp, well-defined melting points; amorphous solids do not.

Crystals

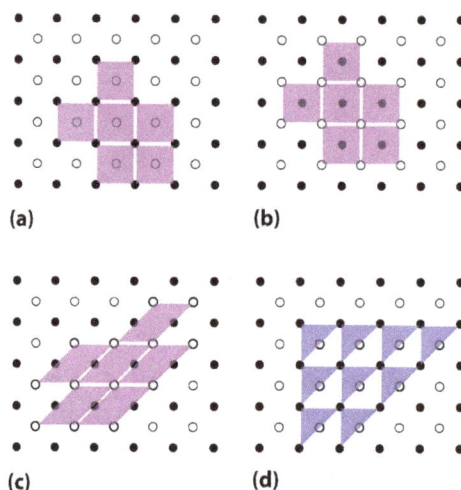

Figure 4: Unit Cells in Two Dimensions. (a–c) Three two-dimensional lattices illustrate the possible choices of the unit cell. The unit cells differ in their relative locations or orientations within the lattice, but they are all valid choices because repeating them in any direction fills the overall pattern of dots. (d) The triangle is not a valid unit cell because repeating it in space fills only half of the space in the pattern.

Because a crystalline solid consists of repeating patterns of its components in three dimensions (a crystal lattice), we can represent the entire crystal by drawing the structure of the smallest identical units that, when stacked together, form the crystal. This basic repeating unit is called a unit cell. For example, the unit cell of a sheet of identical postage stamps is a single stamp, and the unit cell of a stack of bricks is a single brick. In this section, we describe the arrangements of atoms in various unit cells.

Unit cells are easiest to visualize in two dimensions. In many cases, more than one unit cell can be used to represent a given structure, as shown for the Escher drawing in the chapter opener and for a two-dimensional crystal lattice in Figure 4. Usually the smallest unit cell that completely describes the order is chosen. The only requirement for a valid unit cell is that repeating it in space must produce the regular lattice. The concept of unit cells is extended to a three-dimensional lattice in the schematic drawing in Figure 5.

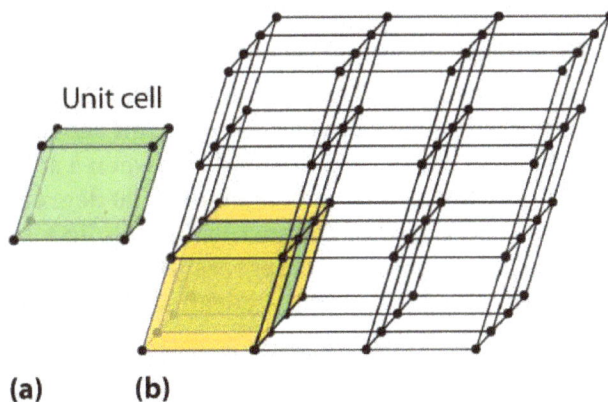

Unit cell

(a)   (b)

Figure 5: Unit Cells in Three Dimensions. These images show (a) a three-dimensional unit cell and (b) the resulting regular three-dimensional lattice.

The Unit Cell

There are seven fundamentally different kinds of unit cells, which differ in the relative lengths of the edges and the angles between them (Figure 6). Each unit cell has six sides, and each side is a parallelogram. We focus primarily on the cubic unit cells, in which all sides have the same length and all angles are 90°, but the concepts that we introduce also apply to substances whose unit cells are not cubic.

If the cubic unit cell consists of eight component atoms, molecules, or ions located at the corners of the cube, then it is called simple cubic. If the unit cell also contains an identical component in the center of the cube, then it is body-centered cubic (bcc). If there are components in the center of each face in addition to those at the corners of the cube, then the unit cell is face-centered cubic (fcc).

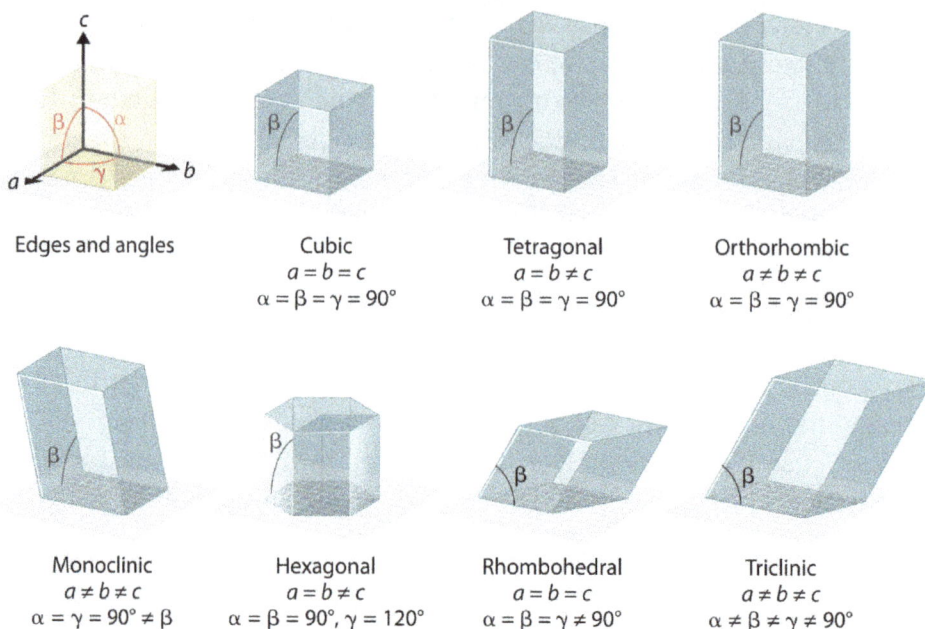

Figure 6: The General Features of the Seven Basic Unit Cells. The lengths of the edges of the unit cells are indicated by a, b, and c, and the angles are defined as follows: α, the angle between b and c; β, the angle between a and c; and γ, the angle between a and b. Cubic: a = b=c, alpha= beta=gamma = 90 degrees. Tetragonal: a=b, alpha = beta = gamma = 90 degrees. Orthorhombic: alpha = beta = gamma = 90 degrees. Monoclinic: alpha = gamma = 90 degrees. Hexagonal: a = b, alpha = beta = 90 degrees, gamma = 120 degrees. Rhombohedral: a = b = c, alpha = beta= gamma. Triclinic: all different.

As indicated in Figure 7, a solid consists of a large number of unit cells arrayed in three dimensions. Any intensive property of the bulk material, such as its density, must therefore also be related to its unit cell. Because density is the mass of substance per unit volume, we can calculate the density of the bulk material from the density of a single unit cell. To do this, we need to know the size of the unit cell (to obtain its volume), the molar mass of its components, and the number of components per unit cell. When we count atoms or ions in a unit cell, however, those lying on a face, an edge, or a corner contribute to more than one unit cell, as shown in Figure 7. For example, an atom that lies on a face of a unit cell is shared by two adjacent unit cells and is therefore counted as 1/2 atom per unit cell. Similarly, an atom that lies on the edge of a unit cell is shared by four adjacent unit cells, so it contributes 1/4 atom to each. An atom at a corner of a unit cell is shared by all eight adjacent unit cells and therefore contributes 1/8 atom to each. The statement that atoms lying on an edge or a corner of a unit cell count as 1/4 or 1/8 atom per unit cell, respectively, is true for all unit cells except the hexagonal one, in which three unit cells share each vertical edge and six share each corner (Figure 7), leading to values of 1/3 and 1/6 atom per unit cell, respectively, for atoms in these positions. In contrast, atoms that lie entirely within a unit cell, such as the atom in the center of a body-centered cubic unit cell, belong to only that one unit cell.

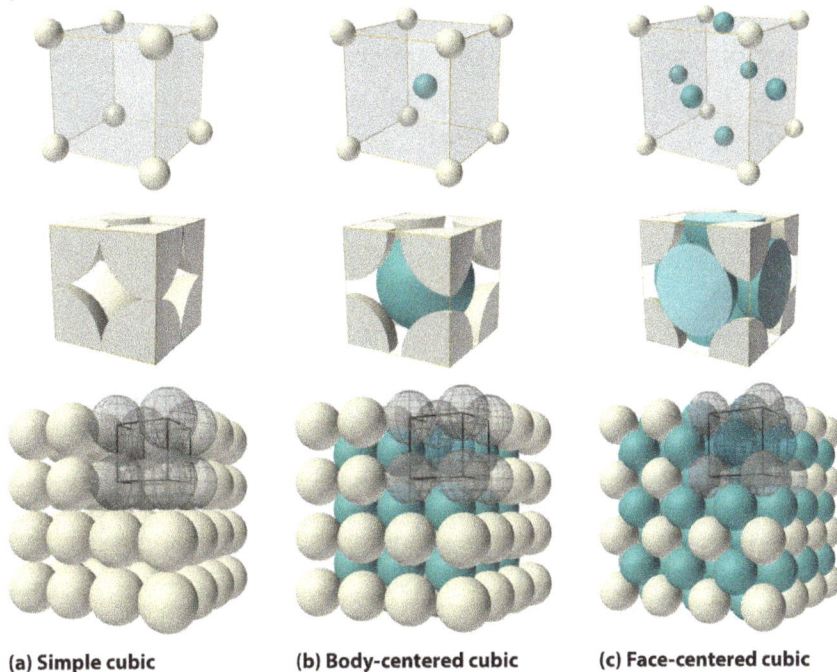

(a) Simple cubic   (b) Body-centered cubic   (c) Face-centered cubic

Figure 7: The Three Kinds of Cubic Unit Cell. For the three kinds of cubic unit cells, simple cubic (a), body-centered cubic (b), and face-centered cubic (c), there are three representations for each: a ball-and-stick model, a space-filling cutaway model that shows the portion of each atom that lies within the unit cell, and an aggregate of several unit cells.Simple cubic is made of eight quarters of spheres. Body center cubic is made of eight quarters of spheers with a shpere in the middle. Face centered cubic is made of eight quarters of spheres and six half spheres.

9.3 IONIC CRYSTALS

Crystals are found everywhere that chemical deposits are located. Ruby crystals are extremely valuable, both because of ruby's beauty and its utility in equipment such as lasers. Some claim that crystals have magical qualities. For others, the "magic" is in the regular structure of the crystal, as the cations and anions line up in a regular order.

Ionic Crystal Structure

Electron dot diagrams show the nature of the electron transfer that takes place between metal and nonmetal atoms. However, ionic compounds do not exist as discrete molecules, as the dot diagrams may suggest. In order to minimize the potential energy of the system, ionic compounds take on the form of an extended three-dimensional array of alternating cations and anions. This maximizes the attractive forces between the oppositely charged ions. The figure below shows two different ways of representing the ionic crystal lattice. A ball and stick model makes it easier to see how individual ions are oriented with respect to one another. A space filling diagram is a more accurate representation of how the ions pack together in the crystal.

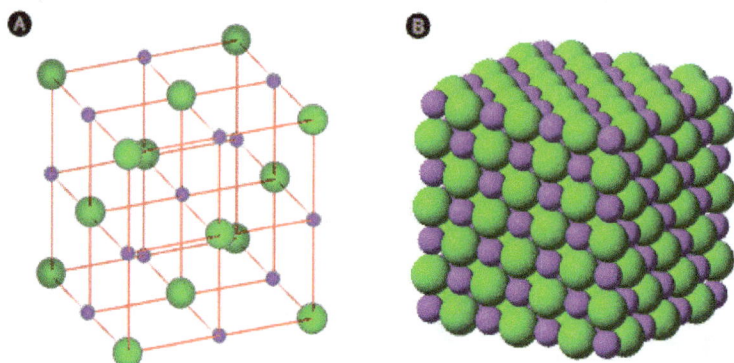

Figure 8: Two models of a sodium chloride crystal are shown. The purple spheres represent the Na$^+$ ions, while the green spheres represent the Cl$^-$ ions. (A) In an expanded view, the distances between ions are exaggerated, more easily showing the coordination numbers of each ion. (B) In a space filling model, the electron clouds of the ions are in contact with each other.

Naturally occurring sodium chloride (halite) does not look at first glance like the neat diagrams shown above. It is only when we use modern techniques to analyze the crystal structure at the atomic level that we can see the true regularity of the organized ions.

Figure 9 Halite crystals.

9.4 COVALENT NETWORK CRYSTALS

Covalent Network Solids are giant covalent substances like diamond, graphite and silicon dioxide (silicon(IV) oxide). This page relates the structures of covalent network solids to the physical properties of the substances.

Diamond

Carbon has an electronic arrangement of 2,4. In diamond, each carbon shares electrons with four other carbon atoms - forming four single bonds.

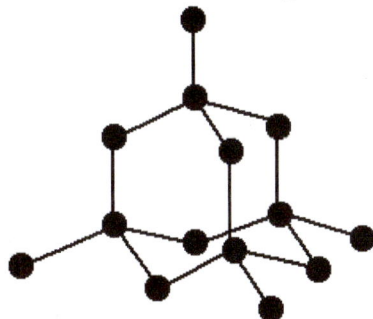

In the diagram some carbon atoms only seem to be forming two bonds (or even one bond), but that's not really the case. We are only showing a small bit of the whole structure. This is a giant covalent structure - it continues on and on in three dimensions. It is not a molecule, because the number of atoms joined up in a real diamond is completely variable - depending on the size of the crystal.

How to draw the structure of diamond

Don't try to be too clever by trying to draw too much of the structure! Learn to draw the diagram given above. Do it in the following stages:

Practice until you can do a reasonable free-hand sketch in about 30 seconds.

Physical Properties of Diamond

- has a very high melting point (almost 4000°C). Very strong carbon-carbon covalent bonds have to be broken throughout the structure before melting occurs.

- is very hard. This is again due to the need to break very strong covalent bonds operating in 3-dimensions.

- doesn't conduct electricity. All the electrons are held tightly between the atoms, and aren't free to move.

- is insoluble in water and organic solvents. There are no possible attractions which could occur between solvent molecules and carbon atoms which could outweigh the attractions between the covalently bound carbon atoms.

Graphite

Graphite has a layer structure which is quite difficult to draw convincingly in three dimensions. The diagram below shows the arrangement of the atoms in each layer, and the way the layers are spaced.

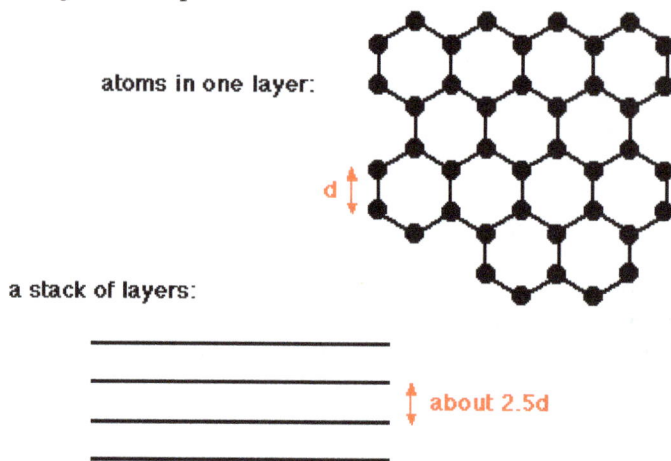

atoms in one layer:

d

a stack of layers:

about 2.5d

Notice that you cannot really draw the side view of the layers to the same scale as the atoms in the layer without one or other part of the diagram being either very spread out or very squashed. In that case, it is important to give some idea of the distances involved. The distance between the layers is about 2.5 times the distance between the atoms within each layer. The layers, of course, extend over huge numbers of atoms - not just the few shown above.

The Bonding in Graphite

Each carbon atom uses three of its electrons to form simple bonds to its three close neighbors. That leaves a fourth electron in the bonding level. These "spare" electrons in each carbon atom become delocalized over the whole of the sheet of atoms in one layer. They are no longer associated directly with any particular atom or pair of atoms, but are free to wander throughout the whole sheet. The important thing is that the delocalized electrons are free to move anywhere within the sheet - each electron is no longer fixed to a particular carbon atom. There is, however, no direct contact between the delocalized electrons in one sheet and those in the neighboring sheets. The atoms within a sheet are held together by strong covalent bonds - stronger, in fact, than in diamond because of the additional bonding caused by the delocalized electrons.

So what holds the sheets together? In graphite you have the ultimate example of van der Waals dispersion forces. As the delocalized electrons move around in the sheet, very large temporary dipoles can be set up which will induce opposite dipoles in the sheets above and below - and so on throughout the whole graphite crystal.

Graphite has a high melting point, similar to that of diamond. In order to melt graphite, it isn't enough to loosen one sheet from another. You have to break the covalent bonding throughout the whole structure. It has a soft, slippery feel, and is used in pencils and as a dry lubricant for things like locks. You can think of graphite rather like a pack of cards - each card is strong, but the cards will slide over each other, or even fall off the pack altogether. When you use a pencil, sheets are rubbed off and stick to the paper. Graphite has a lower density than diamond. This is because of the relatively large amount of space that is "wasted" between the sheets.

Graphite is insoluble in water and organic solvents - for the same reason that diamond is insoluble. Attractions between solvent molecules and carbon atoms will never be strong enough to overcome the strong covalent bonds in graphite. conducts electricity. The delocalized electrons are free to move throughout the sheets. If a piece of graphite is connected into a circuit, electrons can fall off one end of the sheet and be replaced with new ones at the other end.

Silicon dioxide: SiO_2

Silicon dioxide is also known as silica or silicon(IV) oxide has three different crystal forms. The easiest one to remember and draw is based on the diamond structure. Crystalline silicon has the same structure as diamond. To turn it into silicon dioxide, all you need to do is to modify the silicon structure by including some oxygen atoms.

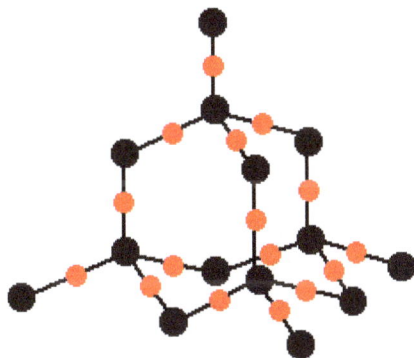

Notice that each silicon atom is bridged to its neighbors by an oxygen atom. Don't forget that this is just a tiny part of a giant structure extending on all 3 dimensions.

Silicon Dioxide has a high melting point - varying depending on what the particular structure is (remember that the structure given is only one of three possible structures), but around 1700°C. Very strong silicon-oxygen covalent bonds have to be broken throughout the structure before melting occurs. Morevoer, it hard due to the need to break the very strong covalent bonds. Silicon Dioxide does not conduct electricity since there aren't any delocalized electrons with all the electrons are held tightly between the atoms, and are not free to move. Silicon Dioxide is insoluble in water and organic solvents. There are no possible attractions which could occur between solvent molecules and the silicon or oxygen atoms which could overcome the covalent bonds in the giant structure.

9.5 METALLIC CRYSTALS AND BAND THEORY

9.5.1 Introduction: An Overview

Metallic crystals represent a fascinating class of materials characterized by their unique electronic structure and physical properties, which are fundamentally governed by the principles of band theory. In solid-state physics, band theory provides a comprehensive framework for understanding the behavior of electrons in solids, elucidating their energy levels and mobility within the crystal lattice. In metallic crystals, the constituent atoms are arranged in a regular, three-dimensional lattice structure. Unlike insulators or semiconductors, metallic crystals have a high density of conduction electrons that are delocalized throughout the crystal lattice. This delocalization arises from the overlapping atomic orbitals, leading to the formation of electron bands or energy bands. Within the context of band theory, the electronic structure of metallic crystals is described by the concept of energy bands and band gaps. In a metallic crystal, the valence band is partially filled with electrons, and there is no distinct energy gap between the valence band and the conduction band. This characteristic feature allows electrons to move freely through the crystal lattice in response to an applied electric field, thus imparting metals with their excellent conductivity and malleability. The conductivity of metallic crystals can be further understood by considering the Fermi level, which represents the highest energy level occupied by electrons at absolute zero temperature. The Fermi level lies within the conduction band for metals, allowing electrons to be easily excited to higher energy states and contribute to electrical conduction. Moreover, the behaviour of electrons in metallic crystals is influenced by various factors such as crystal structure, temperature, and impurities. Changes in temperature can alter the distribution of electrons among energy states, affecting the electrical and thermal conductivity of the material. Similarly, the introduction of impurities or defects can disrupt the regular arrangement of atoms and modify the electronic properties of the metallic crystal. In summary, metallic crystals exhibit unique electronic properties that stem from the delocalization of electrons within the crystal lattice and the absence of a band gap between the valence and conduction bands. The principles of band theory provide valuable insights into the behaviour of electrons in metallic crystals, elucidating their conductivity, thermal properties, and other physical characteristics essential for a wide range of technological applications, including electrical wiring, electronics, and metallurgy.

9.5.2 Bonding mechanisms:

Metallic crystals: Example: iron, copper etc

Each atom loses its valence electrons to a common sea of electrons.

The crystal is made of positive ion centres.

Valence electrons are free to move about the positive ions in the crystal

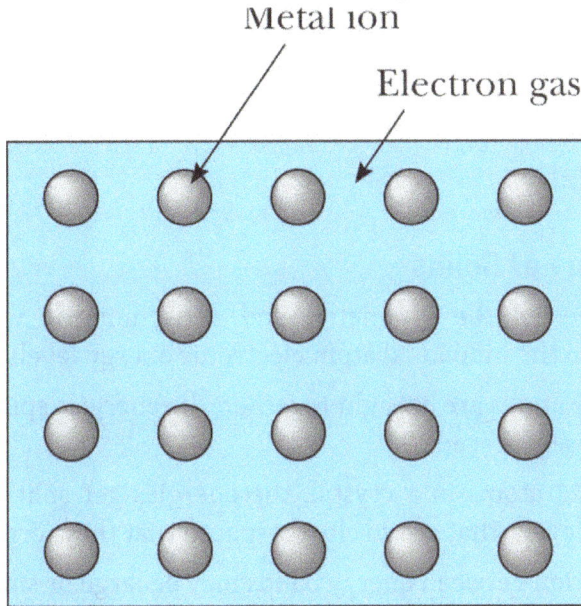

Bonding mechanisms:

Properties of metallic crystals:

- ⊙ Atomic cohesive energy is smaller than for crystals, but still strong.
- ⊙ Electrons are free to move, so electrical conductivity is high (good conductors).
- ⊙ Can absorb and emit visible radiation close to the metal surface.

9.6 BAND THEORY OF SOLIDS

- ⊙ Two identical atoms that are far apart do not interact.
- ⊙ Their electronic energy levels are those of individual atoms.
- ⊙ As the atoms come closer together, the wave functions start overlapping.
- ⊙ The joint wave functions is a symmetric or an anti-symmetric combination of the individual wave functions.

Symmetric combination: $\Psi_+ = \Psi_1 + \Psi_2$

Anti-symmetric combination: $\Psi_- = \Psi_1 + \Psi_2$

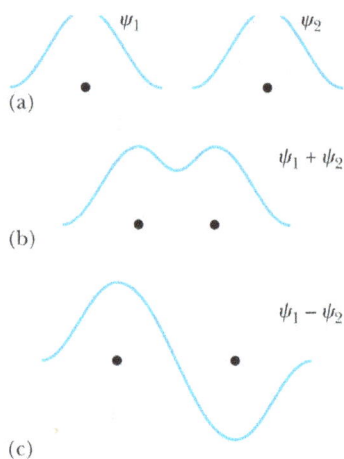

9.6.1 Band Theory of Solids

- The energies of the symmetric and antisymmetric wavefunctions are different. So the individual atom electronic energy levels split into two.

- When many atoms are brought together, the energies split into more levels that are closely spaced.

- For N = 1023 atoms in a crystal, the energies get split into a very large number of levels, that are so closely spaced that they form an energy band.

- The separation between energy bands may be large or small, depending on the atom. Energy bands may overlap.

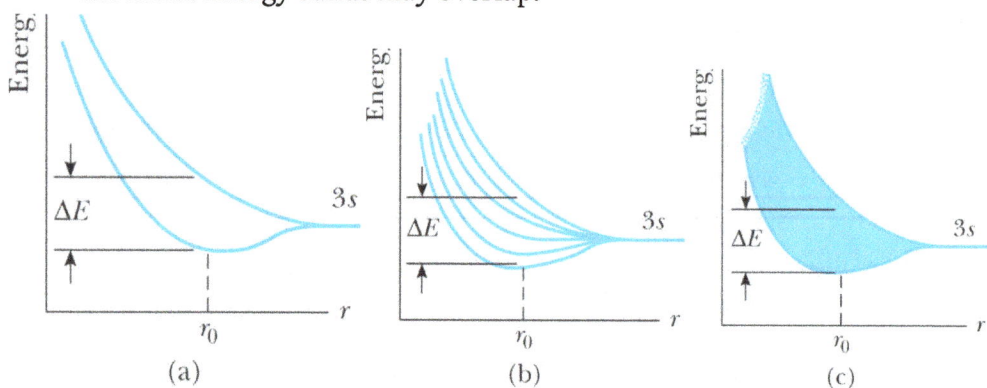

Energy bands

- For N atoms in a crystal each energy band has N energy levels.

- Including the orbital angular momentum and the spin quantum numbers, each band can hold $2(2l+1)N$ electrons.

Example: Sodium: electronic structure: $1s^2 2s^2 2p^6 3s^1$

- The 1s, 2s and 2p bands are full.

- The 3s band has one electron from each atom.
- Total number of electrons in the 3s band: N.
- The 3s band can hold 2N electrons, so it is half full.

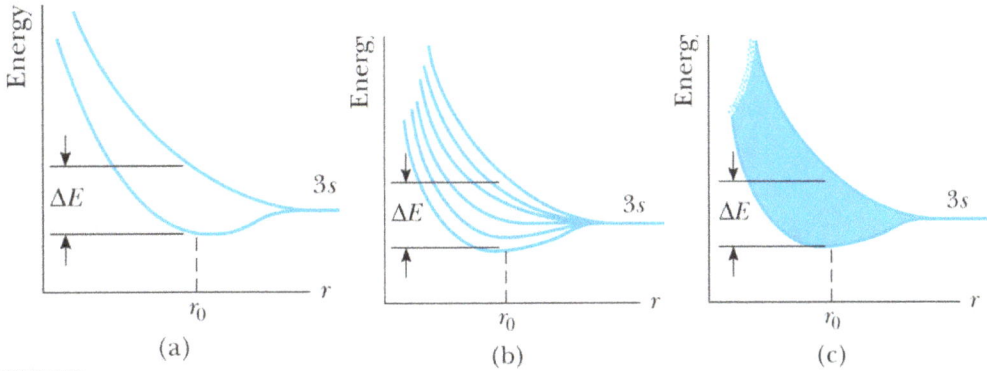

(a) (b) (c)

Energy bands

- Electrons occupying lower lying energy bands are tightly bound to the atom.
- The electrons in the highest energy band participate in conduction.
- The highest occupied energy band is called the **valence band**.
- The lowest energy band with unoccupied states is called the **conduction band**.

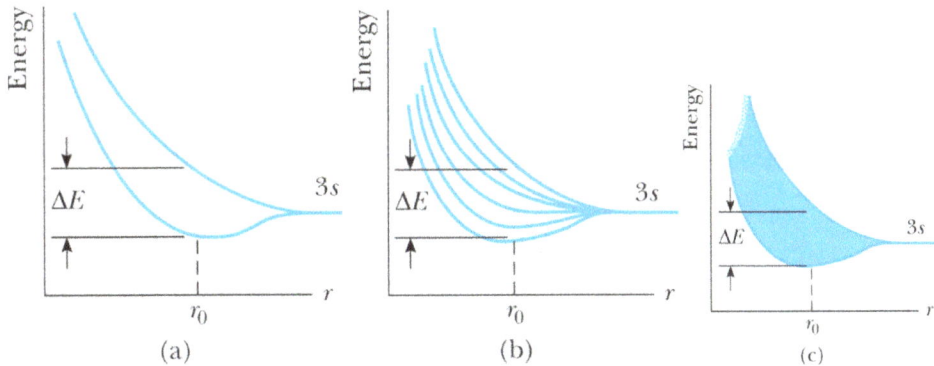

(a) (b) (c)

Metals

- In metals the highest energy band is partially full.
- This band is both the valence band and the conduction band.
- There are many empty energy levels available nearby.
- A small electric field can excite electrons into these empty levels.
- Electrons are thus free to move, hence metals are good conductors.

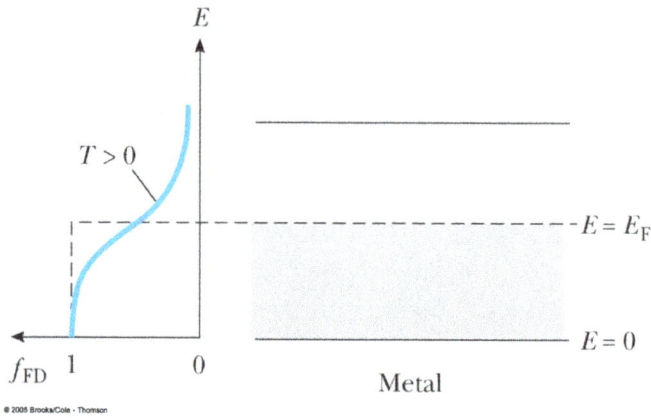

E

$T > 0$

$E = E_F$

$E = 0$

f_{FD} 1   0

Metal

© 2005 Brooks/Cole - Thomson

Metals

◉ In insulators, the valence band is full. The conduction band is empty.

◉ The two bands are separated by a **large energy gap**. The Fermi energy lies between the gap.

◉ A small electric field cannot excite electrons from the valence to the conduction band.

◉ Hence electrons are tightly bound: the material is a **good insulator**

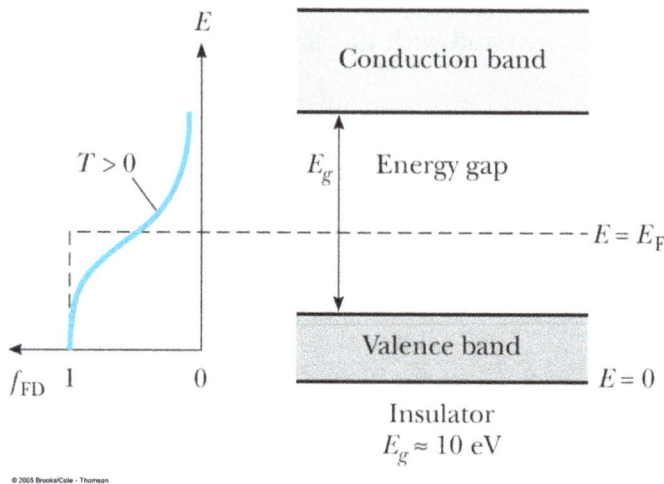

E

Conduction band

$T > 0$

E_g Energy gap

$E = E_F$

Valence band

f_{FD} 1   0 $E = 0$

Insulator
$E_g \approx 10$ eV

© 2005 Brooks/Cole - Thomson

10 | BONDING IN BIOLOGICAL MOLECULES

10.1 BIOLOGICAL MOLECULES: AN OVERVIEW

Bonding in biological macromolecules constitutes the foundational framework upon which life's complexity and diversity are built. These macromolecules, including proteins, nucleic acids, carbohydrates, and lipids, exhibit a remarkable array of structural, functional, and regulatory roles in living organisms, all orchestrated by the intricate interplay of chemical bonds. From the intricate folding patterns of proteins to the double helical structure of DNA, the diverse manifestations of bonding in biological macromolecules underscore the fundamental principles governing cellular processes and molecular interactions. At the core of biological macromolecules lie covalent bonds, where atoms share electron pairs to form stable molecular frameworks. In proteins, for instance, covalent bonds between amino acids create polypeptide chains with specific sequences and spatial arrangements, dictating their three-dimensional structures and functional properties. Similarly, the phosphodiester bonds in nucleic acids link nucleotide monomers to form the backbone of DNA and RNA, encoding genetic information and enabling processes such as replication and transcription.

Complementing covalent bonds are non-covalent interactions, which play pivotal roles in shaping the dynamic behaviour and functionality of biological macromolecules. Hydrogen bonds, electrostatic interactions, van der Waals forces, and hydrophobic interactions contribute to the folding, stability, and assembly of macromolecular structures, facilitating molecular recognition events and driving cellular processes such as signal transduction and molecular transport.

The study of bonding in biological macromolecules transcends disciplinary boundaries, drawing upon insights from chemistry, biology, physics, and computational modelling to elucidate the molecular mechanisms underpinning life's essential processes. Advances in structural biology, spectroscopy, and bioinformatics have revolutionized our ability to probe the intricate architectures and functional landscapes of macromolecular assemblies, offering unprecedented insights into the molecular basis of health and disease. In essence, the exploration of bonding in biological macromolecules represents a journey into the heart of life's molecular machinery, where atoms and molecules choreograph the intricacies of cellular function, organismal development, and evolutionary adaptation. By unravelling the mysteries of macromolecular bonding, scientists seek to unravel the mysteries of life itself, paving the way for transformative discoveries in fields ranging from medicine and biotechnology to agriculture and environmental science.

10.2 PEPTIDE BONDING IN PROTEINS

A peptide bond, also called an eupeptide bond, is a chemical bond that is formed by joining the carboxyl group of one amino acid to the amino group of another. A peptide bond is basically an amide-type of the covalent chemical bond. This bond links two consecutive alpha-amino acids from C1 (carbon number one) of one alpha-amino acid and N_2 (nitrogen number two) of another. This linkage is found along a peptide or protein chain.

During the formation of this bond, there is a release of water (H_2O) molecules. A peptide bond is usually a covalent bond (CO-NH bond), and since the water molecule is eliminated, it is considered a dehydration process. Generally, this process occurs mostly between amino groups. Peptide is a Greek word that means "digested". A peptide is a short polymer of amino acid monomers linked by an amide bond.

10.2.1 Peptide Bond Formation or Synthesis

A peptide bond is formed by a dehydration synthesis or reaction at a molecular level. This reaction is also known as a condensation reaction which usually occurs between amino acids. As depicted in the figure given below, two amino acids bond together to form a peptide bond by dehydration synthesis. During the reaction, one of the amino acids gives a carboxyl group to the reaction and loses a hydroxyl group (hydrogen and oxygen).

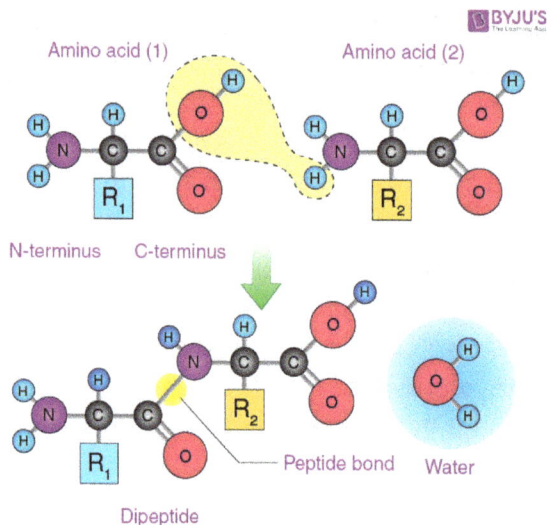

The other amino acid loses hydrogen from the NH2 group. The hydroxyl group is substituted by nitrogen, thus forming a peptide bond. This is one of the primary reasons for peptide bonds being referred to as substituted amide linkages. Both amino acids are covalently bonded to each other.

The newly formed amino acids are also called a dipeptide.

Let's have a look at a simpler diagram depicting the formation of the peptide bond.

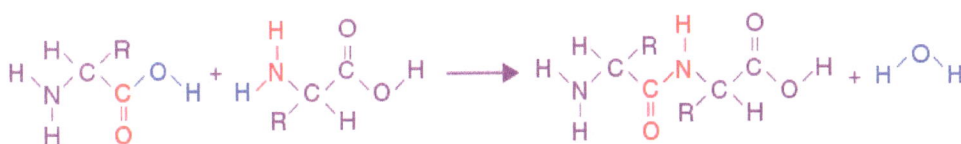

During the reactions that occur, the resulting CO-NH bond is the peptide bond, and the resulting molecule is an amide. The four-atom functional group -C(=O)NH- is called an amide group or a peptide group.

10.2.2 Characteristics of Peptide Bonds

⊙ Peptide bonds are strong with partial double bond character:

➤ They are not broken by heating or high salt concentration.

➤ They can be broken by exposing them to strong acids or bases for a long time at elevated temperature, and also by some specific enzymes (digestive enzymes).

⊙ Peptide bonds are rigid and planar bonds; therefore, they stabilise protein structure.

⊙ Peptide bond contains partial positive charge groups (polar hydrogen atoms of amino groups) and partial negative charge groups (polar oxygen atoms of carboxyl groups).

Different Forms of Peptide Bond

⊙ Dipeptide = contains 2 amino acid units.

⊙ Tripeptide = contains 3 amino acid units.

⊙ Tetrapeptide = contains 4 amino acid units.

⊙ Oligopeptide = contains not more than 10 amino acid units.

⊙ Polypeptide = contains more than 10 amino acid units, up to 100 residues.

⊙ Macropeptides = made up of more than 100 amino acids.

10.2.3 Degradation of Peptide Bond

The degradation of peptide bonds involves a reaction in which the breaking of the peptide bonds between the molecules occurs. Hydrolysis (addition of water) is the reaction used for the degradation of the peptide bond. During the reaction, they will emit Gibbs energy in an amount of 8-16 kJ/mol. But generally, this is a very slow process having a half-life of 350 to 600 years per bond at a temperature of 25°C. Enzymes like proteases are used as the catalysts for this process.

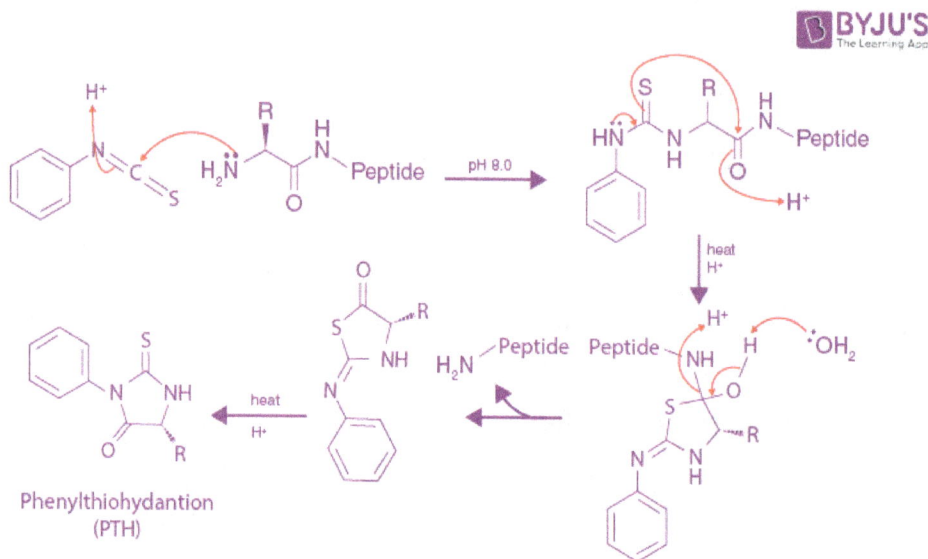

Phenylthiohydantion
(PTH)

10.2.4 Peptide Bond Structure

A peptide bond is a planar, trans and rigid configuration. It also shows a partial double bond character. The coplanarity of the peptide bond denotes the resonance or partial sharing of two pairs of electrons between the amide nitrogen and carboxyl oxygen.

The atoms C, H, N, and O of the peptide bond lie in the same plane, like the hydrogen atom of the amide group and the oxygen atom of the carboxyl group, which are trans to each other.

Linus Pauling and Robert Corey are the scientists who found that the peptide bonds are rigid and planar.

10.2.5 Peptide Bond Features

Some key features of this bond include the following:

◉ Writing of the Peptide Bond Structure

◉ Generally, these bonds are written in a form where free amino acids are at the left and the free carboxyl on the right side. The left side is the

N-terminal residue, and the right side is the C-terminal residue. This amino acid sequence is read from the N-terminal to the C-terminal. And also, protein biosynthesis starts in the same direction.

Representation of the Peptide Bond

Rattlesnake moving representation is used for the peptide bond representation from left to right of the page. The N-terminal residues to its rattle and C-terminal residues are considered as the fangs.

Shorthand to Read the Peptide Bond

The peptide or protein of the amino acid is represented by the 3 letters or one-letter abbreviation.

The Naming of the Peptide Bond

To name the peptides, we should know the suffixes of the amino acids. -ine for glycine, -an for tryptophan, -ate for glutamate, are changed to -yl, except in the case of the C-terminal of the amino acid.

Stereochemistry of the Peptide Bond

We know that every protein is made of simpler units of amino acids with L-configuration. The steric arrangement of the alpha carbon is fixed by that configuration.

Spectra

The wavelength of absorption for a peptide bond is 190–230 nm. With such a reading, it means that the bond is easily susceptible to UV radiation.

Reactions

As a result of its resonance stabilization, a peptide bond is almost unreactive under physiological conditions. It is even less than that of esters. However, in some cases, peptide bonds can undergo chemical reactions, but this is mainly due to an attack from an electronegative atom on the carbonyl carbon. This results in the breaking of the carbonyl double bond and the formation of a tetrahedral intermediate.

10.3 NUCLEIC ACID'S STRUCTURE AND BONDING

Nucleic acids are polymers of nucleotides linked through phosphodiester linkages. Two types of nucleic acids; Deoxyribonucleic acid (DNA) and Ribonucleic acid (RNA) are present in the cell. Among these DNA act as a genetic material and inherit the information from one generation to next. RNA comprises of several types namely messenger RNA (mRNA), ribosomal RNA (rRNA) and transfer RNA (tRNA). RNA play many other roles like rRNA is component of ribosomes while mRNA contains codons that are translated into proteins. tRNA transfers corrects amino acid on the growing polypeptide chain by reading the codons of mRNA. Nucleotides of DNA and RNA are made up of nitrogenous bases, sugar and phosphate. The sugar that is present in nucleic acids is pentose sugar which is of two types; one that present in DNA is 2-deoxy-Dribose and the other present in RNA is D-ribose. Both the pentoses are present as closed five membered ring (β-furanose form) (Figure).

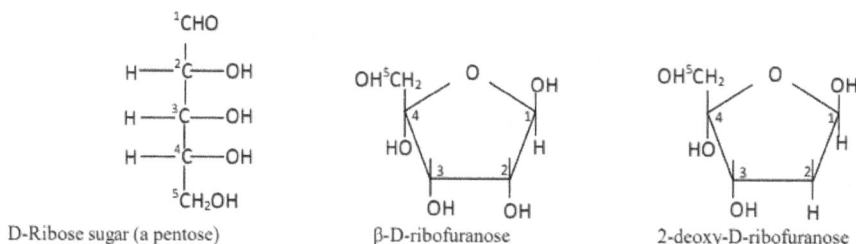

Figure 1: Structure of pentose sugars present in nucleic acids.

Nitrogenous bases are of two types; purines and pyrimidines. The two purine bases of DNA and RNA are adenine (A) and guanine (G). Among pyrimidines cytosine (C) is present in both DNA and RNA, thymine (T) is present in DNA only and uracil (U) is present in RNA only. The structure of five major bases is shown in (Figure 1). The purines and pyrimidens bases contain aromatic ring having conjugated double bonds structures which absorb light a wavelength near 260nm.

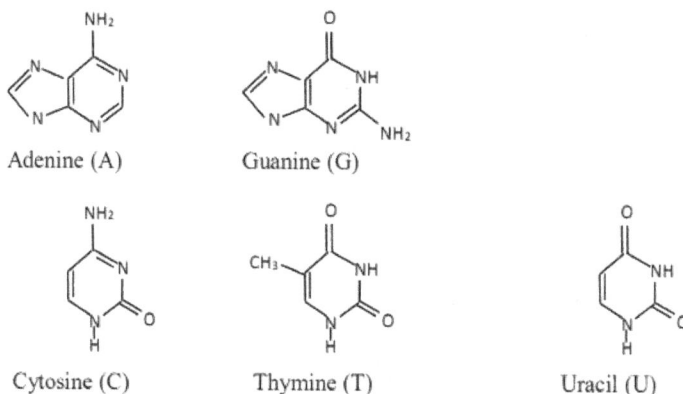

Figure 2 : Structure of nitrogenous bases present in nucleic acids.

The bases, sugar and phosphate linked together to form nucleotide. N-9 (nitrogen 9) of purine bases form an N-β-glycosyl bond to the C-1 (carbon 1) of the pentoses while N-1 of pyrimidine bases form an N-β-glycosyl bond to the C-1 of the pentoses. A phosphate moleculye is esterified to the 5′ carbon of the pentose. The nucleotides in DNA are called deoxyribonucleotides or deoxyribonucleosides-5′- monophosphates form the structural unit of DNA. They are of four types namely; deoxyadenylate (deoxyadenosine-5′-monophosphate), deoxyguanylate (deoxyguanosine-5′-monophosphate), deoxycytidylate (deoxycytidine-5′-monophosphate) and deoxythymidylate (deoxythymine-5′- monophosphate) (Figure).

Deoxyadenylate (deoxyadenosine-5′-monophosphate)

Deoxyguanylate (deoxyguanosine-5′-monophosphate)

Deoxycytidylate (deoxycytidine-5′-monophosphate)

Deoxythymidylate (deoxythymine-5′-monophosphate)

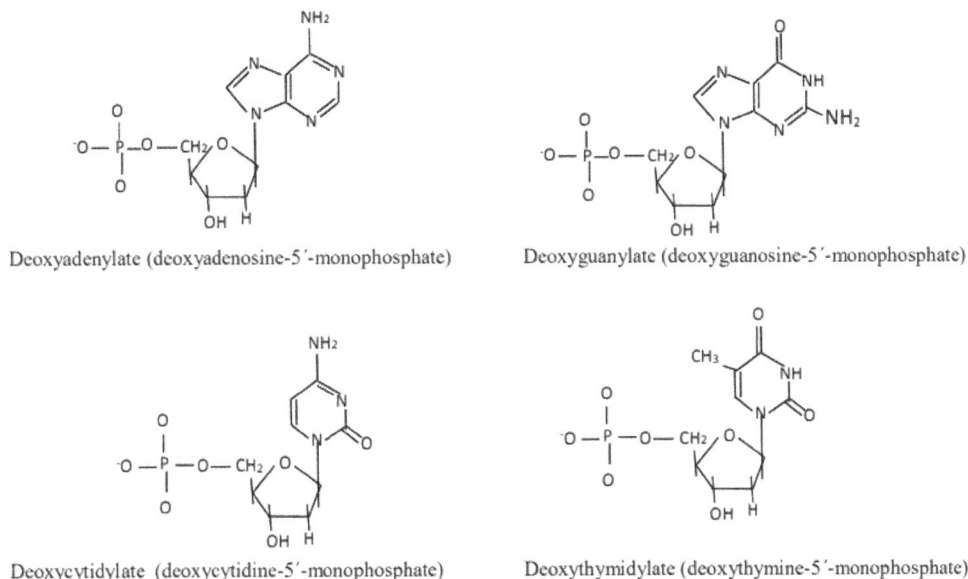

Figure 3: The nucleotides present in DNA.

The nucleotides in RNA are called ribonucleotides or ribonucleotide-5′-monophosphate and are the structural units of RNA. There are four types of ribonucleotides namely; adenylate (adenosine-5′- monophosphate), guanylate (guanine-5′-monophosphate), cytidylate (cytidine-5′-monophosphate) and uridylate (uridine-5′-monophosphate) (Figure).

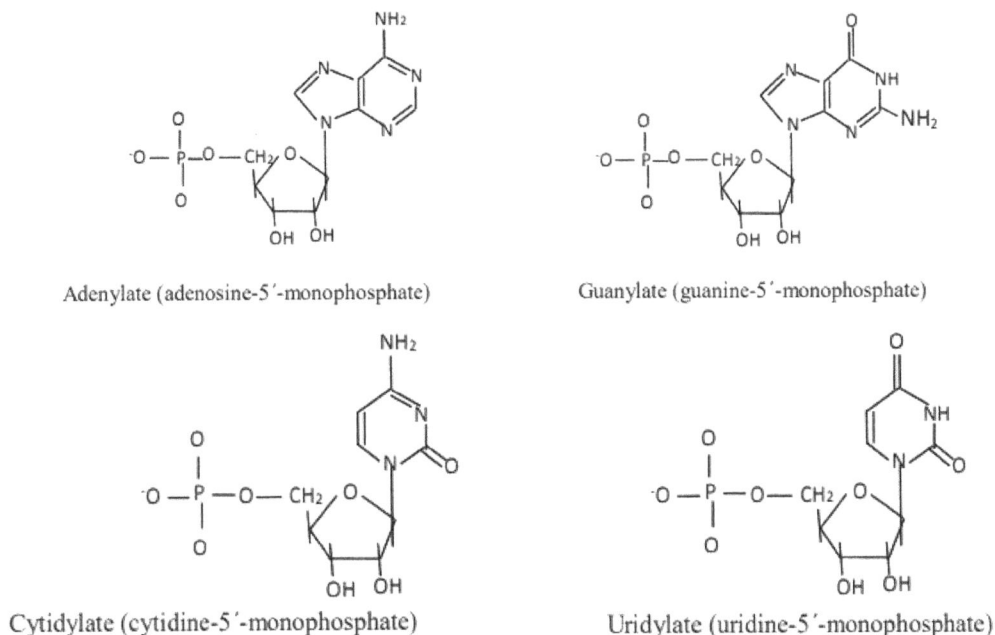

Adenylate (adenosine-5′-monophosphate)

Guanylate (guanine-5′-monophosphate)

Cytidylate (cytidine-5′-monophosphate)

Uridylate (uridine-5′-monophosphate)

Figure 4 : The nucleotides present in RNA.

10.4 PRIMARY STRUCTURE OF NUCLEIC ACIDS

(A.) Primary structure of DNA   (B.) Primary structure of RNA

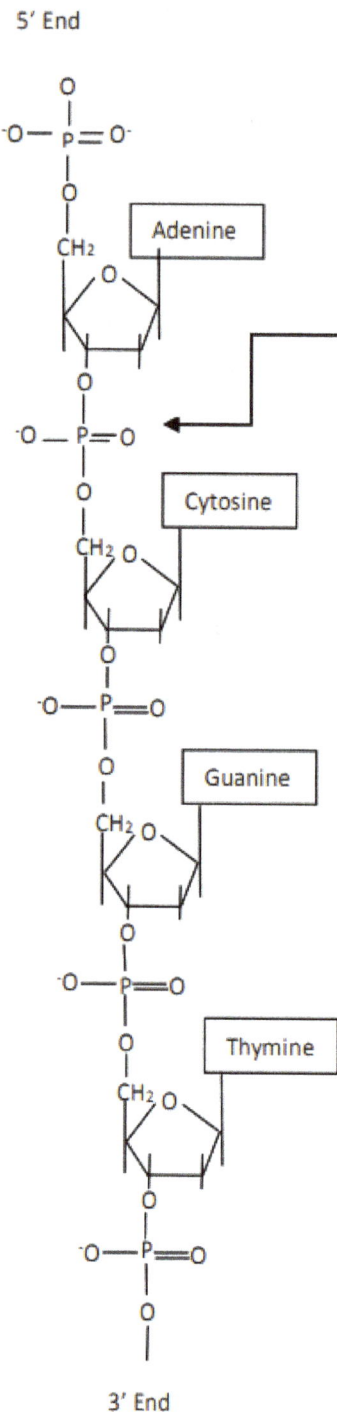

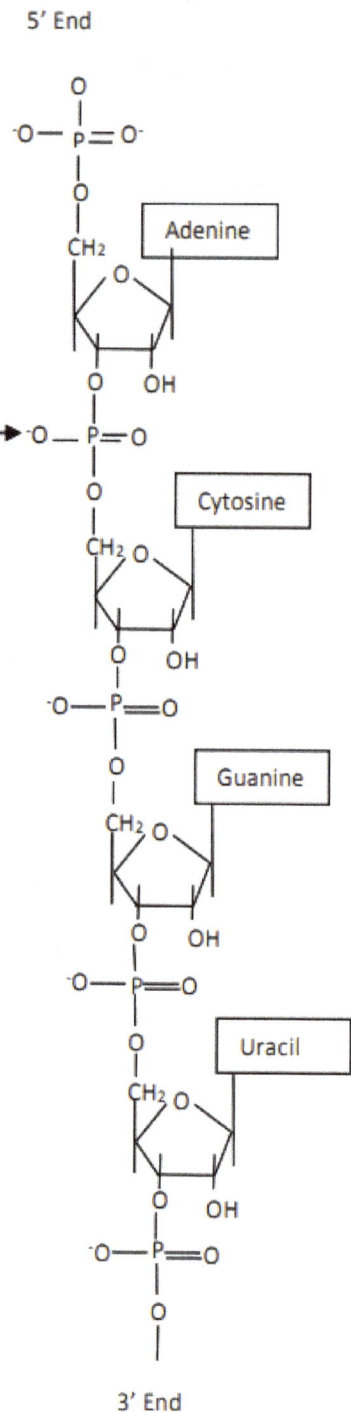

Figure 5 : Primary structure of DNA (A.) and RNA (B.)

Nucleotides of both DNA and RNA are covalently linked by phosphate group in which the 5′ - phosphate group of one nucleotide unit is attached to the 3′ hydroxyl group of the next nucleotide, forming a phosphodiester linkage. Thus the alternating phosphate and pentose residues form the covalent backbones and nitrogenous bases form side groups joined to the backbone at regular intervals (Figure 5). The backbone of both DNA and RNA are hydrophilic as –OH groups of sugar residues form hydrogen bonds with water. All ends of linear polynucleotide chain has a specific polarity i.e. distinct 5′ and 3′ ends. The 5′ end has phosphate group at C-5 ′ of sugar while 3′ end has free -OH group of ribose. The covalent backbone of RNA is hydrolyzed rapidly in alkaline condition. The 2′ -OH group acts nucleophilically on adjacent phosphate group thus hydrolyzing the RNA backbone. Since DNA backbone lacks 2′ -OH, it is stable under similar conditions.

The polynucleotide of upto 50 nucleotides is referred as an oligonucleotide. A larger nucleic acid is called as polynucleotide.

10.4.1 Secondary structure of DNA

James Watson and Francis Crick in 1953 described the three dimensional structure of DNA. It consists of two polynucleotide chains of DNA wound around the same axis to form right handed double helix (Figure 6). The two strands are oriented antiparallel i.e. their 3′ and 5′ phosphodiester bonds run in opposite directions. The sugar and phosphate backbone forms the outer circumference of the double helix therefore exposed to polar environment having water. The bases of both the strands are stacked inside the core of the helix makes it hydrophobic. The rings of bases are planer and perpendicular to the helix axis. The surface of DNA double helix has two grooves, one is called major groove and other is minor groove. Within the helix each nucleotide base of one strand makes hydrogen bonds in the same plane with a base of the other strand. A of one strand forms two hydrogen bonds with T of the other strand and vice versa and G on the one strand forms three hydrogen bonds with C on the other strand vice versa. So, two hydrogen bonds can form between A=T and T=A, and three hydrogen bonds can form between G≡C and C≡G. Because of these three hydrogen bonds separation of DNA helices needs more energy when ratio of GC to AT bp is higher.

The two antiparallel polynucleotide chains of DNA helix are not identical but they are complementary to each other. Whenever A occurs in one strand, T will be present in the other strand and vice versa similarly whenever G occurs in one chain, C will be present in the other strand and vice versa. The distance between the stacked bases double helix is 3.4 A° . The distance for making one complete helix turn is 34 A° . The number of base pairs in each complete turn of double helix is 10.5. The double helix of DNA is held together by two forces. One is hydrogen bonding between the complementary bases pairs and second is the base stacking interaction.

bonds. The special nature of these bonds determines much of the shape of the more complex compound, largely because these linkages inhibit the rotation of specific molecules toward each other.

Starch, cellulose, and glycogen are homopolysaccharides because they contain only one type of monomer, glucose. There are heteropolysaccharides which contain more than one monomer, such as the peptidoglycans in bacterial cell walls. Mostly there are not more than two different kinds of monomer in carbohydrate polymers. We will find that carbohydrates are less differentiated in their polymer structure than proteins.

10.5.3 Starch

Starch occurs as granules in plant cells. It breaks down to become a large number of alpha-D-glucose molecules. Starch has alpha-glycosidic bonds to link glucose molecules. The preferred conformation of amylose, the simplest form of starch, is a helix. The energy that holds together the helical shape and the glycosidic linkages comes free when it is broken down in plants or in the digestive tract of animals and humans. Its role is to be a major energy source in the living world. Enzymes in plant, animal and human organisms can easily break down the alpha-glycosidic linkages of the starch helix to yield glucose, and glucose can be broken down further to yield energy. The alpha -linkage between the glucose molecules in starch also determines its function as an energy storage compound.

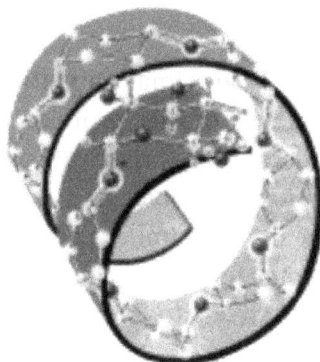

Figure 7: Starch (from Campbell, 1999)

10.5.4 Cellulose

Cellulose is the main component of the cell wall of plants. Cellulose is formed from -D-glucose with -glycosidic linkages. The -glycosidic linkages of cellulose allow for additional hydrogen bonding between linear polysaccharide chains. This results in a strong planar shape which can neither be broken down easily in plants nor in the digestive tract of humans and many animals. The typical -linkages of cellulose, and the possibility of hydrogen bonding which this allows, make cellulose a structural carbohydrate in plants. The cell wall around the cell membrane of plants consists mainly of cellulose, giving plants their stability. Woody plants contain more cellulose.

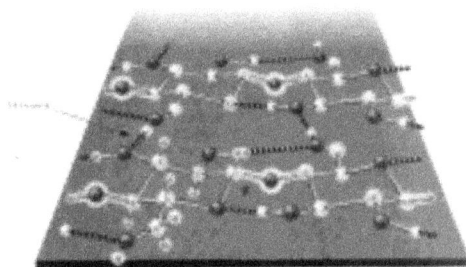

Figure 8: Cellulose (from Campbell, 1999)

Further compounds with -glycosidic linkages

Structural carbohydrates in non-plants have amino acids or contain amino acid sequences as monomers. Plant cell walls contain relatively little protein or peptide. Carbohydrates with -glycosidic linkages can be found in some invertebrates such as insects, shrimp, or lobster. Their exoskeleton contains chitin, which is the polymer of Nacetyl--D-glucosamine, a monosaccharide with an amine group added onto the sugar. In chitin, individual strands are held together by hydrogen bonds as in cellulose. Accordingly, chitin has a structural function. It is also found in the cell walls of yeasts, fungi and algae. -Glycosidic linkages also connect the two amine-group-containing monomers in bacterial cell walls: N-acetyl--D-glucosamine and N-acetylmuramic acid. The strands are crosslinked by amino acid residues, forming a peptidoglycan. Peptidoglycans form a strong structure that is the target of certain antibiotic agents.

10.5.5 Glycogen

Glycogen is found in granules in certain types of cells in animals and humans, like liver and muscle cells, but not normally in heart and brain cells in the human organism. -Glycosidic linkages connect the glucose molecules in glycogen. As in starch, the alpha -glycosidic linkages allow for glycogen's function in energy storage because glucose can readily be cleaved off.

10.5.6 Proteins

When proteins are hydrolyzed this results in a large number of amino acids. Unlike the monomers of polysaccharides, the amino acids in proteins are of many different forms. Twenty different amino acids are found in human protein in varying quantities and combinations. They are linked together by peptide bonds to form the primary structure of proteins. Peptide bonds in proteins are also specialized covalent bonds, like the glycosidic bonds in carbohydrates. And, like glycosidic linkages, peptide bonds inhibit rotation of specific molecules in the amino acids around each other and therefore play a role in the final shape of proteins.

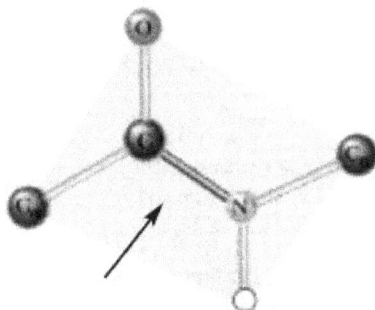

Figure 9: The peptide bond (from Campbell, 1999)

However, the final shape of proteins is not exclusively determined by the sequence of amino acids and the peptide bonds linking them together (the protein's primary structure of covalent bonds). The conformation of proteins is also subject to intricate folding processes connected to different types of bonds such as hydrogen bonds and disulfide bonds. The primary structure of proteins, though, determines their ability to form a secondary and tertiary structure, which is required for proteins to be biologically active in the organism. The secondary structure of proteins is based on the hydrogen-bonded arrangement of the protein's amino acid backbone. It is responsible for -helix and -pleated sheet sections in the protein chain. Hydrogen bonds are important in the final structure of collagen, a structural protein.

The tertiary structure of proteins adds to their actual three-dimensional structure with the help of covalent disulfide bonds between sulfide-containing amino acid side chains, hydrogen bonding between amino acid side chains, electrostatic forces of attraction and hydrophobic interactions. Proteins can exhibit a rod-like fibrous or a compact globular conformation, depending not only on the bonding forces mentioned above but also on the conditions under which the protein is formed.

Proteins can also have a quaternary structure, which involves several different polypeptide chains. The bonds involved to hold this protein structure together are noncovalent. The conformation of a protein is specific to it and determines its function and its functional ability.

10.5.7 Lipids

Acetyl-CoA can be considered an important common component of lipids. In the group of lipids, fatty acids and cholesterol are both ultimately synthesized from acetyl-CoA. The open chain lipids contain one or more of the fatty acids, and the fused ring lipids, the steroids, are conversions of cholesterol. Fatty acids are also oxidized to acetyl-CoA. Cholesterol derivatives are not broken down in the human body but are excreted. Lipids have the tendency to form clusters in the watery milieu of the body because they have long nonpolar tails which are hydrophobic. The hydrophobic fatty acid tails are "hidden" inside the cluster, sequestered from water, and at the periphery of the cluster the water-soluble hydrophilic side, connected to the head group, is exposed. Hydrophobic interactions occur spontaneously in aqueous surroundings. In terms of thermodynamics, this type of interaction does not require added energy when hydrophobic side chains or tails are present in the watery milieu of the body. This is in contrast to the types of bonding described earlier, which do require extra energy to occur.

Lipid clusters can take on either micelle or membrane-like forms to allow sequestering of their hydrophobic portions from watery surroundings. The micelle form occurs, for instance, when lipids are taken up into and transported in the body. Micelles have a single layer of lipids in their structure. They get a high degree of complexity in low-density lipoprotein (LDL) particles, for instance, in which a mosaic of cholesterol and phospholipids bound to a protein (apoprotein B-100) forms the outer structure around many molecules of cholesteryl esters. LDL particles play an important role in the transport of cholesterol in the blood stream. The triacylglycerols (triglycerides) are the energy storage form of lipids and accumulate as fat globules in the cells of adipose tissue. All lipids except for triglycerides can be found as components of membranes. Membranes: Lipids are necessary for the formation of membranes because of their hydrophobic property, and the consequent clustering that occurs. Membranes are bilayers of lipids. They divide water into compartments, which is essential for the functioning of all organisms. In single-cell organisms it makes the formation of organelles possible and separates the organism from its (mostly aqueous) environment.

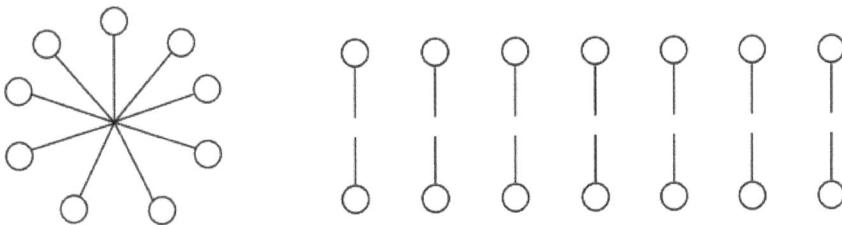

The micelle form   The membrame form

Figure 10: The micelle and membrane forms of lipids. (indicates the hydrophilic backbone of the lipid, indicates the hydrophobic tail of the lipid). The micelle can be large and have an inner space that is sequestered from water in which lipids can be transported, as in LDL particles. Membranes are actually also round structures and enclose an inner space, the interior of cells or cell compartments.

Membranes are semi-permeable mainly as a result of the presence of proteins, which function as channels in between the lipids. This allows for the transportation of compounds across the membrane and ensures the connection between the watery milieu inside the membrane with that outside the membrane. Semi-permeable membranes allow for the possibility of a different intracellular milieu from the environment. This makes the single cell into an organism. In multi-cell organisms, membranes make differentiated functioning possible. Cells can have different functions and yet have the same extracellular environment. Metabolism mainly occurs in the intracellular milieu. The extracellular milieu has an important role in transportation of metabolic substances and connecting the cells in the organism to form a whole. In vertebrates, the separation of intracellular and extra-cellular milieu makes functions such as the contraction of muscles and conduction of electrical impulses in the nervous system possible.

The presence of the fused-ring lipid, cholesterol, which is rather rigid in its structure, and of unsaturated fatty acids, which have kinks in their tail portion, influences the fluidity of membranes in opposite ways. The membranes of prokaryotes (such as bacteria) are the most supple of all since they hardly contain steroids. Plant membranes contain phytosterols (a steroid similar to cholesterol) and have less fluidity. The many unsaturated fatty acids in plant membranes make them more fluid than membranes in animals and humans, which contain cholesterol. Lipids (possibly except for the triacylglycerols) occur in specific structured forms in organisms. Lipid structures such as micelles and membranes could perhaps be seen as the polymer form of lipids. The defining interaction in these structures is the spontaneously occurring hydrophobic interaction, supported by weak and changeable van der Waals bonds. Hydrophobic interactions induce the relative immobility of the lipid components toward each other. However, all components of membranes are in flux, as the seemingly constant shape of organisms is always in movement.

11 | UNCONVENTIONAL AND EXOTIC BONDING

11.1 HYDROGEN- HYDROGEN BONDING

Hydrogen bonding is a type of intermolecular force that occurs between a hydrogen atom bonded to an electronegative atom (such as oxygen, nitrogen, or fluorine) and another electronegative atom in a nearby molecule. The electronegative atom pulls electron density away from the hydrogen atom, leaving the hydrogen atom partially positively charged. This partial positive charge allows the hydrogen atom to attract the lone pair of electrons on the electronegative atom of another molecule, forming a relatively strong dipole-dipole interaction. Hydrogen bonds are responsible for many important properties of substances, including the high boiling points of water and the secondary structure of proteins. Some common examples of hydrogen bonding include the bonds between water molecules, which contribute to the unique properties of water such as its high surface tension, cohesion, and ability to dissolve many substances. Hydrogen bonding also plays a crucial role in biological systems. For instance, it helps stabilize the structure of DNA and proteins, influencing their properties and functions. In summary, hydrogen bonding is a specific type of intermolecular force that occurs between a hydrogen atom bonded to an electronegative atom and another electronegative atom in a neighboring molecule. It is an important phenomenon in understanding the properties and behavior of many substances, both in the laboratory and in nature.

Hydrogen bond is a weak bond formed between a H atom and highly electronegative atom like O, N or F either of the same molecule or of a different molecule but to which it is not directly attached. Moreover the H atom itself should be attached to a highly electronegative atom like O, N or F.

Example: Consider the hydrogen fluorine bond in hydrogen fluoride. This bond in a polar covalent bond is which fluorine is strong electronegative element. As a result the fluorine acquires a partial negative charge and hydrogen acquires a partial positive charge.

$$\overset{+\delta}{H} - \overset{-\delta}{F}$$

(Hydrogen Flouride)

The lone pair on the fluorine atom in another molecule of hydrogen fluoride will attract the positive charge on hydrogen in a molecule of hydrogen fluoride electrostatically. This bond between hydrogen and fluorine of different molecules is known as hydrogen bond. This new type of linkage is represented by dotted lines

$$\overset{+\delta}{H} - \overset{-\delta}{F} ....\overset{+\delta}{H} - \overset{-\delta}{F} ....\overset{+\delta}{H} - \overset{-\delta}{F}$$

Hydrogen Bond

Conditions For Hydrogen Bonding

◉ The molecule must contain a highly electronegative atom linked to hydrogen atom.

◉ The size of electronegative atom should be small these conditions are met in only by F, O and N atoms.

Although Cl has the same electronegativity as nitrogen, it does not form effective hydrogen bond. This is because of its larger size than that of N with the result its electrostatic attractions are weak

Types of Hydrogen Bonding

Generally the hydrogen bonds are classified into two types.

Intermolecular Hydrogen Bonding

In such types of hydrogen bonding the two or more than two molecules of same or different compounds combine together to give a polymeric aggregate.

For example,

(i)
$$\overset{+\delta}{H} - \overset{-\delta}{O} .... \overset{+\delta}{H} - \overset{-\delta}{O} .... \overset{+\delta}{H} - \overset{-\delta}{O}$$
with $+\delta H$ below each O

(ii)
$$\overset{+\delta}{H} - \overset{-\delta}{N} .... \overset{+\delta}{H} - \overset{-\delta}{N} .... \overset{+\delta}{H} - \overset{-\delta}{N}$$
with $+\delta H$ above and below each N

(iii)
H – C with O–H .... O and O .... H – O groups, C – H

(iv)

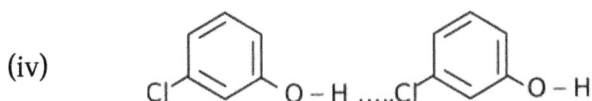

Intramolecular Hydrogen Bonding

In this type, hydrogen bonding occurs within two atoms of the same molecule. This type of hydrogen bonding is commonly known as chelation and frequently occurs in organic compounds.

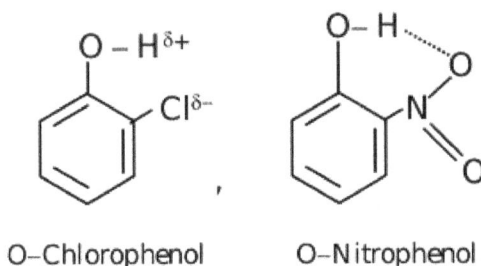

O–Chlorophenol   O–Nitrophenol

Hydrogen bonding has got a very pronounced effect on certain properties of the molecules.

State of the Substance

H_2O exists in liquid state whereas H_2S in gaseous state because hydrogen bonding exist in water and no H-bonding exists in H_2S.

Solubility

The organic compounds like alkane, alkenes and alkynes are insoluble in water due to absence of H-bonding whereas alcohols, organic acids, amines are soluble in water due to H-bonding.

Boiling Point

High boiling and melting points of NH_3, H_2O and HI in comparison to hydrides of other elements of V, VI and VII groups to which N, O and F belong respectively are due to hydrogen bonding.

11.2 BORON CLUSTERS AND BORON NITROGEN COMPOUNDS

Boron clusters and boron nitrogen compounds are fascinating areas of study in chemistry with numerous applications in materials science, catalysis, and medicine.

11.2.1 Boron Clusters: Boron clusters are molecular species composed of boron atoms. They often form intricate three-dimensional structures, which can range from simple arrangements to highly complex architectures. Boron clusters are typically classified based on their structural motifs, which can include boranes, carboranes, metallacarboranes, and polyhedral boranes.

- ◉ *Boranes:* Boranes are the simplest boron clusters, consisting of boron atoms bonded together with hydrogen atoms. The most famous example is diborane (B_2H_6), which has a bridged structure.

- ◉ *Carboranes:* Carboranes are boron clusters that contain both boron and carbon atoms, along with hydrogen atoms. They often exhibit unique properties due to their hybrid nature.

- ◉ *Metallacarboranes:* These boron clusters incorporate transition metal atoms into the carborane framework, adding additional functionality and stability.

- ◉ *Polyhedral Boranes:* Polyhedral boranes are highly symmetric boron clusters, often resembling geometric shapes such as icosahedra or dodecahedra. Examples include decaborane ($B_{10}H_{14}$) and closo-dodecaborate ($B_{12}H_{12}^{2-}$).

11.2.2 Boron Nitrogen Compounds: Boron nitrogen compounds, also known as boron nitrides, are compounds that contain both boron and nitrogen atoms. Similar to carbon, boron forms stable compounds with nitrogen, leading to a variety of interesting materials.

- ◉ *Boron Nitride (BN):* Boron nitride exists in several forms, including hexagonal boron nitride (h-BN) and cubic boron nitride (c-BN). Hexagonal boron nitride resembles graphite in structure and is often referred to as "white graphite" due to its similar appearance. Cubic boron nitride, on the other hand, is a superhard material often used as an abrasive and in cutting tools.

- ◉ *Borazine (Inorganic Benzene):* Borazine ($B_3N_3H_6$) is an inorganic compound with a structure analogous to benzene but with alternating boron and nitrogen atoms in place of carbon. It's known as "inorganic benzene" due to its structural resemblance.

- ◉ *Boron Nitride Nanotubes (BNNTs):* Boron nitride nanotubes are nanostructures analogous to carbon nanotubes but composed of boron and nitrogen atoms. They exhibit excellent thermal and chemical stability and have potential applications in nanotechnology and materials science.

11.3 TRANSITION METAL COMPLEXES AND COORDINATION BANDS

Transition metal complexes represent a captivating realm within chemistry, where the intricate interplay between transition metals and ligands yields compounds of immense diversity and functionality. At the heart of these complexes lies the concept of coordination chemistry, where transition metals, characterized by their partially filled d orbitals, exhibit a remarkable propensity to form coordination bonds with surrounding ligands. These ligands, often organic or inorganic molecules possessing lone pairs of electrons, orchestrate a symphony of interactions, engendering structures and properties that defy simplicity. The coordination bonds formed between transition metals and ligands create intricate coordination spheres, where the coordination number, geometry, and electronic configuration dictate the compound's behavior. The study of transition metal complexes transcends mere synthesis, delving into catalysis, materials science, bioinorganic chemistry, and beyond. Embedded within these complexes lie coordination bands, which underpin their electronic and optical properties, influencing phenomena ranging from coloration to magnetic behavior. Exploring the realm of transition metal complexes and coordination bands unveils a rich tapestry of molecular architecture and functionality, offering insights into fundamental chemical principles and avenues for innovative applications.

11.3.1 REACTIONS OF TRANSITION METAL COMPLEXES

The reactions of transition metal complexes may be divided as;

- ◉ Substitution reactions at the metal centre (it involve both SN1 and SN2 types of reaction mechanisms);
- ◉ Oxidation-reduction reactions; and
- ◉ Reactions of ligands that do not change the attachments to the central metal atom. The different types of reactions are individually discussed herewith in detail.

Substitution reactions in octahedral complexes:

Substitution reactions are those in which the replacement of one ligand by another incoming ligand takes place. Since ligands are nucleophilic in nature hence these replacement/substitution reactions are known as nucleophilic substitution reactions (SN). However, there are some reactions known in which the central metal ion is replaced by another metal ion, these reactions are known as electrophilic substitution reactions (SE).

Consider a general substitution reaction in which one of the ligand (X) is replaced by another ligand (Y).

$$[MX_6] + Y \rightarrow [MX_5Y] + X$$

Although there are several mechanisms by which nucleophilic substitution reactions (SN) may occur, however, most of these reactions follow the two types of mechanism frequently. These are:

1. Nucleophilic substitution reactions of first order kinetics (SN1)
2. Nucleophilic substitution reactions of second order kinetics (S 2)

a). Nucleophilic substitution reactions of first order kinetics (S 1): According to this mechanism the reacting complex first undergoes dissociation of one of the metal-ligand bond at which ligand substitution has to take place. This dissociation of metal-ligand bond leads the formation of a penta-coordinated intermediate. The penta coordinated intermediate then readily bonded with the new ligand Y to result the substituted product. The overall mechanism may be shown as:

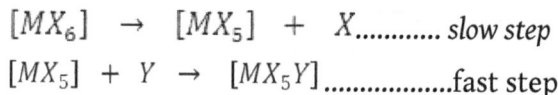

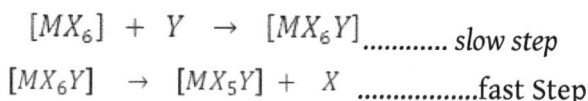

$$[MX_6] \rightarrow [MX_5] + X \dots\dots\dots\text{ slow step}$$
$$[MX_5] + Y \rightarrow [MX_5Y] \dots\dots\dots\text{fast step}$$

In above mechanism in first step the metal ligand bond of complex [MX6] get dissociates to give a penta coordinated intermediate [MX5], this step is a slow step. In second step the penta coordinated intermediate reacts with new ligand to give the substituted product. The second step is a fast step. According the chemical kinetics the slowest step of the multistep reaction is the rate determining step; therefore, in above reaction the rate of reaction depends only on the concentration of [MX6]. The rate law for such reaction may be represented as:

Rate = k [MX6]

Since the rate of above reaction depends only on the concentration of one molecule hence this reaction is also known as **Unimolecular Nucleophilic Substitution Reaction i.e. S 1 mechanism.** As the reaction involves the dissociation of metal ligand bond (which is the key step of this mechanism) hence this reaction is also known as **dissociative nucleophilic substitution reaction (D).**

b). Nucleophilic substitution reactions of second order kinetics (SN2): According to this mechanism the incoming ligand first adds to the reacting complex and formed a hepta- coordinated intermediate. The hepta-coordinated intermediate then readily decomposes to result the substituted product. The overall mechanism may be shown as:

$$[MX_6] + Y \rightarrow [MX_6Y] \dots\dots\dots\text{ slow step}$$
$$[MX_6Y] \rightarrow [MX_5Y] + X \dots\dots\dots\text{fast Step}$$

In above mechanism in first step the reacting complex [MX6] get associated with incoming

ligand to give a hepta coordinated intermediate [MX6Y], this step is a slow step. In second step the hepta coordinated intermediate loses a ligand to give the substituted product. The second step is a fast step. According the chemical kinetics the slowest step of the multistep reaction is the rate determining step; therefore, in above reaction the rate of reaction depends on the concentration of [MX6] and Y. The rate law for such reaction may be represented as:

Rate = k [MX6][Y]

Since the rate of above reaction depends only on the concentration of two molecule hence this reaction is also known as **Bimolecular Nucleophilic Substitution Reaction i.e. SN mechanism.** As the reaction involves the association of reacting complex with incoming ligand, hence this reaction is also known as **associative nucleophilic substitution reaction (A).**

Apart from these two mechanism substitution reaction also occurs through interchanging of ligands. An **interchange substitution reaction** (I) is a direct replacement of the ligand of reacting complex with the incoming ligand and it does not proceeds via formation of an intermediate. The reaction proceeds through the formation of a single transition state leading to the conversion of reactants to products.

$$[MX_6] + Y \rightarrow [MX_5Y] + X$$

If the above reaction is irreversible the interchange substitution reaction mechanism was found to follow second order kinetics. The rate of this reaction depends upon the concentration of reacting complex and the incoming ligand.

$$\text{Rate} = k\,[MX6][Y]$$

However, if the interchange substitution reaction is reversible, then according to the steady state approximation approach the rate of this reaction was found to have pseudo first order reaction kinetics. This will be discussed separately in the kinetics of substitution reactions section.

Types of intermediates formed during unimolecular and biomolecular nucleophilic substitution reaction:

Let us first consider the *unimolecular nucleophilic substitution reaction (SN)* *or dissociative substitution reaction (D)* in which a complex [MX6] is a reacting complex and Y is the incoming nucleophile. Since, the first step of this reaction is a dissociation of metal- ligand bond which results the formation of an intermediate. In dissociative SN1 reaction, penta-coordinated intermediate is formed which may have two types of geometries *i.e.* a) square pyramidal (SP), and b) trigonal bipyramidal (TBP).

The intermediate [MX5] with square pyramidal geometry is formed by the dissociation of metal-ligand (M-X) bond causing the least disturbance in original octahedral geometry. The intermediate [MX5] is then attacked by incoming ligand Y to give [MX5Y] as product. However, the intermediate [MX5] with trigonal bipyramidal geometry is formed by the dissociation of metal-ligand (M-X) bond and the remaining five M-X bonds of intermediate [MX5] immediately adjust the bond angles to produce an intermediate with trigonal bipyramidal geometry which is then attacked by incoming ligand Y to give the product. Since, the trigonal bipyramidal (TBP) intermediate involves the shifting of at least two metal-ligand bonds; however, no such shifting is required during the formation of a square pyramidal (SP) intermediate. It has been observed that under normal conditions the square pyramidal geometry is relatively more stable than the trigonal bipyramidal geometry, therefore, the unimolecular nucleophilic substitution reaction (SN1) or dissociative substitution reaction (D) proceeds through a square pyramidal (SP) intermediate under normal conditions.

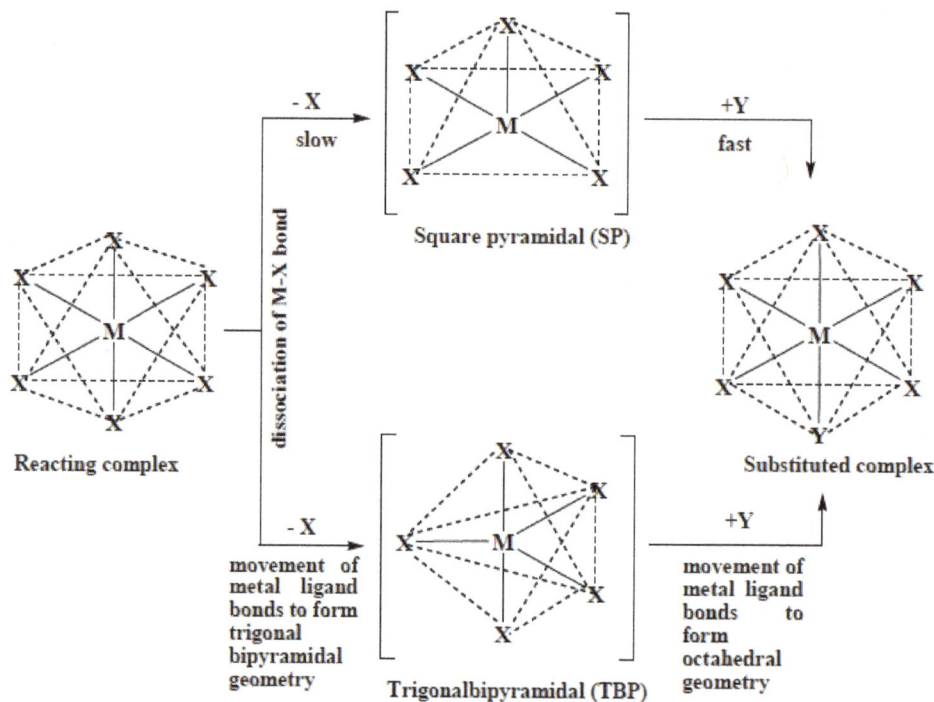

Figure 1: Mechanism of Substitution reaction in octahedral complexes

Secondly, if the above transformation occurs through **bimolecular nucleophilic substitution reaction (SN) or Associative substitution reaction (A),** where the association of reacting complex with incoming ligand takes place in first step which leads the formation of seven coordinated intermediate. The seven coordinated intermediate may also have two possible geometries depending upon the nature of association of reacting complex with incoming ligand viz. a) pentagonal bipyramidal (PBP), and b) octahedral wedge (OW). The pentagonal bipyramidal intermediate is formed when incoming ligand Y approaches the metal ion through one of the edges of the octahedron. The formation of pentagonal bipyramidal (PBP) intermediate requires the shifting of at least four metal-ligand bonds to accommodate the incoming ligand. This shifting of metal-ligand bonds reduces the bond angle and it brings the bonded pair of electrons of ligand-metal-ligand bond angle closer to each other hence the ligand-ligand repulsion and the bond pair – bond pair repulsion increases the energy of the system. That makes the pentagonal bipyramidal intermediate less stable. However, the formation of octahedral wedge (OW) intermediate involves when incoming ligand Y approaches from the middle of one of the triangular faces of octahedron. As the incoming ligand Y approaches towards central metal M the outgoing ligand X starts moving towards the middle of another triangular face, this results the formation of octahedral wedge intermediate.

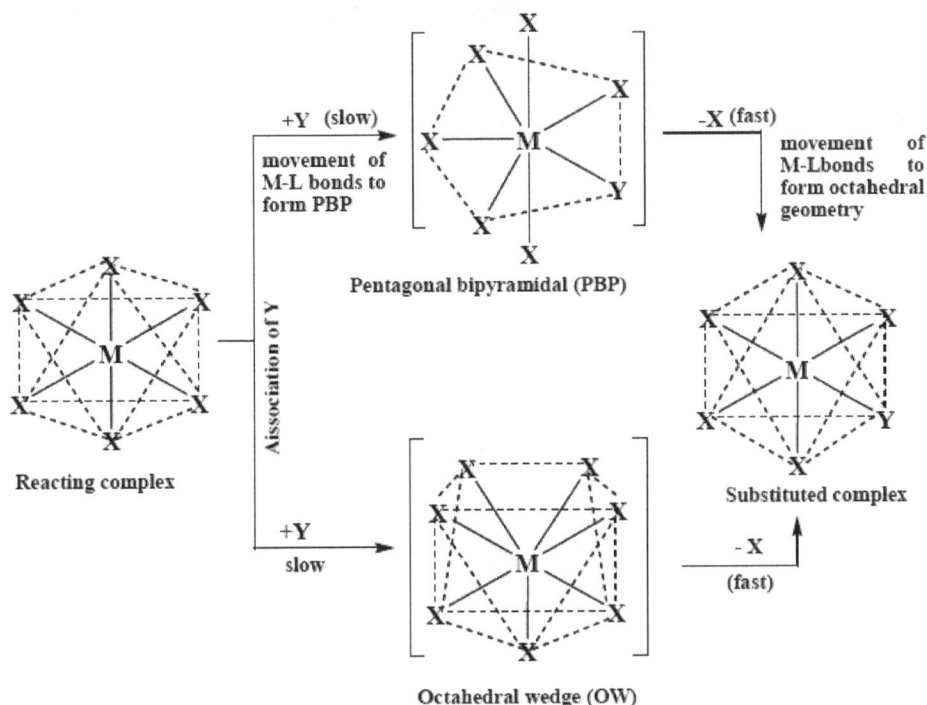

Figure 2: Mechanism of substitution reaction in octahedral complexes via associative path. The formation of octahedral wedge intermediate requires comparatively less shifting of metal ligand bonds thus there is not much alteration in ligand-metal-ligand bond angles. Therefore, in octahedral wedge (OW) intermediate the ligand –ligand repulsion and bond pair-bond pair repulsions are comparatively less than that of observed in pentagonal bipyramidal (PBP) intermediate. Hence, the octahedral wedge (OW) intermediate is more stable than pentagonal bipyramidal (PBP) intermediate.

11.4 COORDINATION BONDS

The hemoglobin in your blood, the chlorophyll in green plants, vitamin B-12, and the catalyst used in the manufacture of polyethylene all contain coordination compounds. Ions of the metals, especially the transition metals, are likely to form complexes. Many of these compounds are highly colored. In the remainder of this chapter, we will consider the structure and bonding of these remarkable compounds. Remember that in most main group element compounds, the valence electrons of the isolated atoms combine to form chemical bonds that satisfy the octet rule. For instance, the four valence electrons of carbon overlap with electrons from four hydrogen atoms to form CH4. The one valence electron leaves sodium and adds to the seven valence electrons of chlorine to form the ionic formula unit NaCl. Transition metals do not normally bond in this fashion. They primarily form coordinate covalent bonds, a form of the Lewis acid-base interaction in which both of the electrons in the bond are contributed by a donor (Lewis base) to an electron acceptor (Lewis acid). The Lewis acid in coordination complexes, often called a central metal ion (or atom), is often a transition metal or inner transition metal, although main group elements can also form coordination compounds. The Lewis base donors, called ligands, can be a wide variety of chemicals—atoms, molecules, or ions. The only requirement is that they have one or more electron pairs, which can be donated to the central metal. Most often, this involves a donor atom with a lone pair of electrons that can form a coordinate bond to the metal.

Figure 3. Metal ions that contain partially filled d subshell usually form colored complex ions; ions with empty d subshell (d0) or with filled d subshells (d^{10}) usually form colorless complexes. This figure shows, from left to right, solutions containing [M(H2O)6]n+ ions with M = Sc3+(d0), Cr3+(d3), Co2+(d7), Ni2+(d8), Cu2+(d9), and Zn2+(d10). (credit: Sahar Atwa)

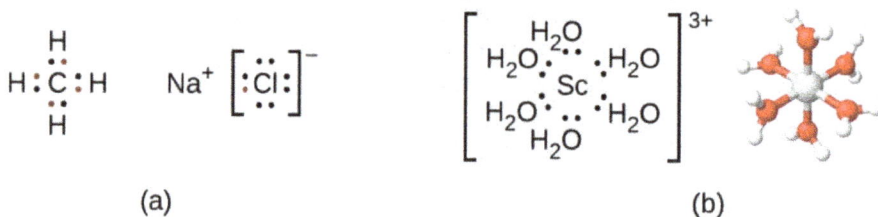

(a)   (b)

Figure 4. (a) Covalent bonds involve the sharing of electrons, and ionic bonds involve the transferring of electrons associated with each bonding atom, as indicated by the colored electrons. (b) However, coordinate covalent bonds involve electrons from a Lewis base being donated to a metal center. The lone pairs from six water molecules form bonds to the scandium ion to form an octahedral complex. (Only the donated pairs are shown.)

The coordination sphere consists of the central metal ion or atom plus its attached ligands. Brackets in a formula enclose the coordination sphere; species outside the brackets are not part of the coordination sphere. The coordination number of the central metal ion or atom is the number of donor atoms bonded to it. The coordination number for the silver ion in $[Ag(NH_3)_2]^+$ is two. For the copper(II) ion in $[CuCl_4]^{2-}$, the coordination number is four, whereas for the cobalt(II) ion in $[Co(H_2O)_6]^{2+}$ the coordination number is six. Each of these ligands is monodentate, from the Greek for "one toothed," meaning that they connect with the central metal through only one atom. In this case, the number of ligands and the coordination number are equal.

Many other ligands coordinate to the metal in more complex fashions. Bidentate ligands are those in which two atoms coordinate to the metal center. For example, ethylenediamine (en, $H_2NCH_2CH_2NH_2$) contains two nitrogen atoms, each of which has a lone pair and can serve as a Lewis base. Both of the atoms can coordinate to a single metal center. In the complex $[Co(en)_3]^{3+}$, there are three bidentate en ligands, and the coordination number of the cobalt(III) ion is six. The most common coordination numbers are two, four, and six, but examples of all coordination numbers from 1 to 15 are known.

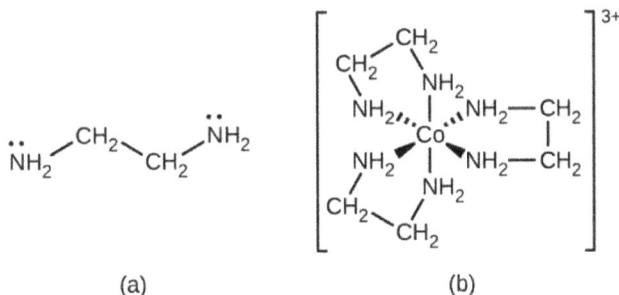

(a)   (b)

Figure 5. (a) The ethylenediamine (en) ligand contains two atoms with lone pairs that can coordinate to the metal center. (b) The cobalt(III) complex $[Co(en)_3]^{3+}$ contains three of these ligands, each forming two bonds to the cobalt ion.

Any ligand that bonds to a central metal ion by more than one donor atom is a polydentate ligand (or "many teeth"), because it can bite into the metal center with more than one bond. The term chelate (pronounced "KEY-late") from the Greek for "claw" is also used to describe this type of interaction. Many polydentate ligands are chelating ligands, and a complex consisting of one or more of these ligands and a central metal is a chelate. A chelating ligand is also known as a chelating agent. A chelating ligand holds the metal ion rather like a crab's claw would hold a marble. Figure 4 showed one example of a chelate. The heme complex in hemoglobin is another important example Figure 5. It contains a polydentate ligand with four donor atoms that coordinate to iron.

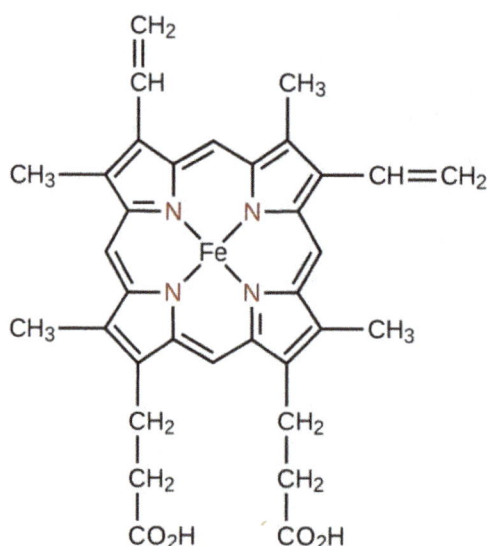

Figure 6. The single ligand heme contains four nitrogen atoms that coordinate to iron in hemoglobin to form a chelate.

Polydentate ligands are sometimes identified with prefixes that indicate the number of donor atoms in the ligand. As we have seen, ligands with one donor atom, such as NH_3, Cl^-, and H_2O, are monodentate ligands. Ligands with two donor groups are bidentate ligands. Ethylenediamine, $H_2NCH_2CH_2NH_2$, and the anion of the acid glycine, $NH_2CH2CO-2NH_2CH_2CO_2-$ are examples of bidentate ligands. Tridentate ligands, tetradentate ligands, pentadentate ligands, and hexadentate ligands contain three, four, five, and six donor atoms, respectively. The ligand in heme (Figure 6) is a tetradentate ligand.

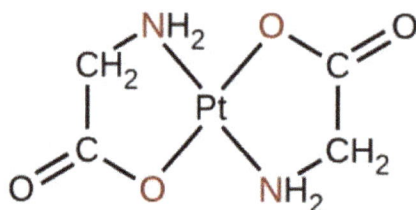

Figure 7. Each of the anionic ligands shown attaches in a bidentate fashion to platinum(II), with both a nitrogen and oxygen atom coordinating to the metal.

11.5 Π- Π STACKING AND OTHER NON-CLASSICAL BONDS

"π-π stacking" refers to a type of non-covalent interaction between aromatic rings. This interaction occurs when two aromatic rings are close enough that their π electron clouds can interact. The π electron clouds are regions of electron density above and below the plane of the aromatic ring, resulting from the delocalized π electrons in the ring structure.

π-π stacking interactions are prevalent in aromatic compounds such as benzene, pyridine, and various other organic molecules containing conjugated π systems. These interactions contribute to the stability and structure of many organic molecules and are important in fields such as chemistry, biology, and materials science.

The strength of π-π stacking interactions can vary depending on factors such as the distance between the aromatic rings, the orientation of the rings relative to each other, and the presence of substituents on the aromatic rings. In general, closer proximity and optimal alignment between the π electron clouds result in stronger π-π stacking interactions.

Apart from π-π stacking, other types of non-classical bonds and interactions exist, including:

- ◉ **Hydrogen Bonds:** These are electrostatic attractions between a hydrogen atom bonded to an electronegative atom (such as nitrogen, oxygen, or fluorine) and another electronegative atom. Hydrogen bonds are crucial in stabilizing the structure of molecules like water and DNA.

- ◉ **Van der Waals Interactions:** These are weak, short-range forces that arise from induced dipoles in molecules. They include London dispersion forces, dipole-dipole interactions, and dipole-induced dipole interactions. Van der Waals forces contribute to the stability of molecular structures and the interactions between molecules.

- ◉ **Hydrophobic Interactions:** These interactions occur between nonpolar molecules in an aqueous environment. Hydrophobic molecules tend to aggregate together to minimize their contact with water molecules, leading to the formation of structures such as lipid bilayers in cell membranes.

- ◉ **Metal Coordination Bonds:** In metal coordination compounds, metal ions are bound to one or more ligands through coordinate covalent bonds. These bonds form between the metal ion and electron-rich regions of the ligands, such as lone pairs of electrons or π electron clouds.

- ◉ **π-Backbonding:** In organometallic chemistry, π-backbonding refers to the donation of electron density from filled d-orbitals of a metal atom into the empty π* antibonding orbitals of an adjacent ligand. This interaction helps stabilize metal-ligand complexes and can influence their reactivity and electronic properties.

Understanding these non-classical bonds and interactions is crucial for elucidating the structure, stability, and reactivity of molecules in various chemical systems.

12 ADVANCING IN CHEMICAL BONDING RESEARCH

12.1 INTRODUCTION: AN OVERVIEW

Chemical bonding stands as one of the foundational pillars of chemistry, intricately woven into the fabric of matter and the interactions that define its properties. As researchers delve deeper into the realm of chemical bonding, they unlock not only the secrets of molecular structure but also the potential for groundbreaking advancements in materials science, drug design, and nanotechnology. At its essence, chemical bonding elucidates the fundamental forces that bind atoms together, shaping molecules and influencing their behavior in diverse environments. From the covalent bonds that underpin organic compounds to the electrostatic attractions governing the formation of ionic solids, the study of chemical bonding transcends disciplinary boundaries, merging physics, chemistry, and materials science in a quest for understanding and innovation. With ever-evolving experimental techniques and computational methods, researchers are poised to unravel the complexities of bonding interactions, forging new frontiers in synthesis, catalysis, and molecular design. As we embark on this journey of discovery, the quest to unravel the mysteries of chemical bonding not only deepens our understanding of the natural world but also paves the way for transformative technologies that promise to shape the future of science and society.

12.2 CUTTING-EDGE TECHNIQUES IN STUDYING CHEMICAL BONDS

The electronic configurations of the elements, as specified in the previous chapter, apply in principle only to isolated atoms – atoms separated by distances over which no mutual interactions of their electronic orbitals can occur (infinite distance). This condition is never met in condensed phases (i.e., liquids and solids); it is only encountered in high vacua where atoms move over long distances without mutual interaction. Under normal conditions, particularly in the mentioned condensed phases, atoms are separated over distances controlled, in essence, by the dimension of their respective outermost occupied electronic orbitals. Whenever the outer electron shells of two or more atoms come in contact with each other (overlap to any extent), the potential for interaction (reaction) exists. Spontaneous reaction (consisting of a rearrangement of the electronic orbitals and/or actual transfer of electrons from one atom to another) will take place whenever such a rearrangement results in a lower energy configuration. This means that the driving force for mutual interaction and rearrangement of electronic configurations is in most instances (but not always) manifested through the release of heat (a form of energy) to the environment. [Typical case: Mg + 1/2 O2 → Reaction Product (MgO) + (light + heat) i.e. energy.] Since the same amount of energy must be supplied to the system if the original state – i.e., separation of the species – is to be reestablished, the reaction partners are "bonded together" by comparable energies. The "strength" of the bonds can obviously vary from system to system with the nature of the electronic rearrangement. Even the inert gases, particularly the heavier ones like xenon, are capable of forming associations with other atoms. Sometimes we find that two atoms assume a more stable state by sharing electrons; at other times, an atom may transfer electrons to another atom in order to achieve a greater stability. In still other instances, the rearrangement may simply be an orbital distortion or an internal charge redistribution. In either event the mutual benefit that accrues is the formation of what is commonly called a chemical bond. Through these bonds atoms combine with each other to form very different kinds of particles referred to as molecules and ions.

12.2.1 . NATURE OF CHEMICAL BONDS

In order for a chemical bond to be formed between two atoms, there must be a net decrease in the energy of the system (the two atoms): the ions or molecules produced by electronic rearrangements must be in a lower energy state than the atoms were prior to interaction, prior to bond formation. Since atoms of each of the elements have different electronic structures, the variety of possible chemical bonds (differing from each other in at least some small way) is considerable and is even further increased by the effects of neighboring atoms on the bond under consideration. The modes of bond formation can be categorized into two basic types, each representing a type of bonding. The bonding types are called electrovalent (or ionic) bonding and covalent bonding. Electrovalent bonding arises from complete transfer of one or more electrons from one atom to another; covalent bonding arises from the sharing of two or more electrons between atoms. Since these models represent the limiting cases, we can anticipate that most real bonds will fall between these two extremes.

Before discussing these models in detail, it is appropriate to consider the relationships between the electronic structures of atoms and their chemical reactivity. The inert gases (Group VIII) are the most stable elements with regard to bond formation, i.e. toward electronic rearrangements. It is therefore useful to examine the reasons for their stability. Inert gases all have electronic structures consisting of filled subshells. For all but helium the outer (or valence) shell contains eight electrons, with filled s and p sublevels (ns^2p^6). The electronic structure of helium is $1s^2$, which is equivalent to the structure of the other inert gases since there is no 1p sublevel. Inert gases have high ionization energies because each electron in the sublevel of highest energy is poorly screened from the nucleus by other electrons in its same sublevel. Each electron "sees" relatively high positive charge on the nucleus, and a large amount of energy is required to remove it from the atom. Inert gases have very low electron affinity because any added electron must enter a significantly higher energy level. We find, therefore, that the electronic structures of inert gases are particularly resistant to changes by either loss or gain of electrons and, further, that atoms of other elements with fewer or more electrons than inert gas configuration tend to gain or lose electrons, respectively, to achieve such inert gas structure.

12.2. 2. ELECTROVALENT (IONIC) BONDING

An electrovalent bond is formed by the transfer of one or more electrons from one atom to another. Consider first atoms that have electronic structures differing from an inert gas structure by only a few, (1, 2 or 3) electrons. These include the representative elements of Groups I, II and III in the Periodic Table, which have respectively 1, 2 and 3 electrons more than a neighboring inert gas, and the representative elements of Groups V, VI and VII, which have respectively 3, 2 and 1 electrons less than a neighboring inert gas (fig. 1).

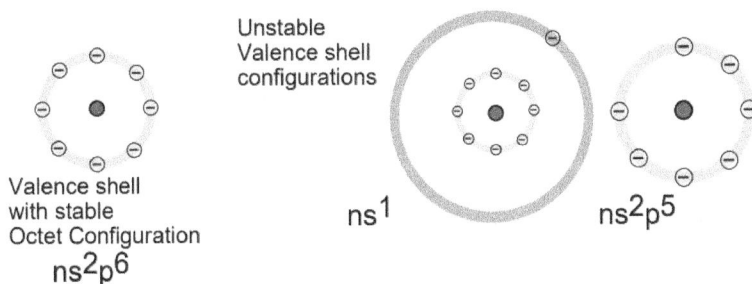

Figure 1: Stable and Unstable Valence Shell Configurations.

The elements of Groups I, II and III can form the electronic structure of an inert gas by losing their outer 1, 2 and 3 (valence) electrons. (The resulting species are positively charged ions.)

In a similar electron transfer which, however, involves the acquisition of electrons in the outer valence levels, elements of Groups V, VI and VII form an inert gas electronic structure (by formation of negatively charged ions).

It is through the electron transfer between an electron-losing element and an electron-gaining element that compounds are formed which involve electrostatic attraction (electrovalent bonds) of oppositely charged species called ions.

$$\text{Na} \bullet \xrightarrow{1e^-} + \text{F} \rightarrow [\text{Na}]^+ + [\text{F}]^-$$

$$\text{Mg} : \xrightarrow{2e^-} + 2\text{F} \rightarrow [\text{Mg}]^{++} + 2[\text{F}]^-$$

(In the notation [Na•], the dot indicates the outermost electron which is in excess of the rare gas configuration. It is referred to as a valence electron.)

Elements immediately following the inert gases (in the horizontal columns of the Periodic Table) lose electrons, and those immediately preceding the inert gases gain electrons on interaction. The resulting compounds are called electrovalent; the valence number (charge on the ion) of a particular element when it forms an electrovalent compound is given by the number of electrons lost or gained in changing from the atomic to the ionic state.

The stoichiometric formula of an electrovalent compound reflects the ratio (usually very simple) of positive to negative ions that gives a neutral aggregate. Hence, the ions Na+ and F– form a compound whose formula is NaF because these ions are singly charged and are present in the compound in a one-to-one ratio. Magnesium nitride, composed of Mg^{2+} and N^{3-}, has the formula Mg_3N_2 because this composition represents electroneutrality.

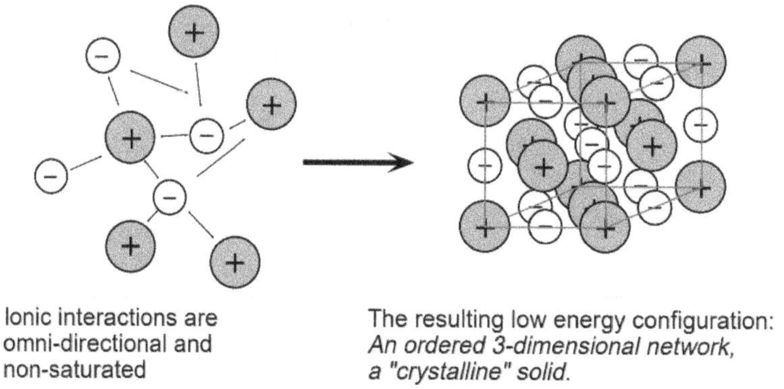

Ionic interactions are omni-directional and non-saturated

The resulting low energy configuration: *An ordered 3-dimensional network, a "crystalline" solid.*

Figure 2: Formation of a crystalline, ordered body (a SOLID), a direct consequence of ion interaction in conjunction with energy minimization.

The electrovalent bond is the result of electrostatic attraction between ions of opposite charge. This attractive force accounts for the stability of these compounds, typified by NaF, LiCl, CaO, and KCl. The ions individually possess the electronic structures of neighboring inert gases; their residual charge arises from an imbalance in the number of electrons and protons in their structures. Isolated ions and simple isolated pairs of ions, as represented by the formula NaCl, exist only in the gaseous state. Their electrostatic forces are active in all directions; they attract oppositely charged species and thus can form regular arrays, resulting in ordered lattice structures, i.e. the solid state (fig. 2). Even in the liquid state and in solutions (where disruptive thermal forces reach values close to that of the attractive electrostatic bonding forces) attraction between ions and with other species remains effective.

12.2.4 Energetics of Ionic Bonding

Ionic bonding is the simplest type of chemical bonding to visualize, since it is totally (or almost totally) electrostatic in nature. The principle of the energetics of ionic bond formation is realized when considering the formation of NaCl (our common salt) from its constituents Na (metal) and Cl2 (chlorine gas). Formally, this reaction is:

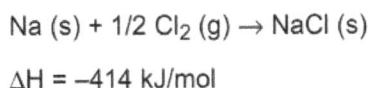

$$Na\ (s) + 1/2\ Cl_2\ (g) \rightarrow NaCl\ (s)$$

$$\Delta H = -414\ kJ/mol$$

The equation as written indicates that 1 mole sodium reacts with 1/2 mole chlorine (Cl_2) under formation of 1 mole (ionically bonded) sodium chloride; this reaction is accompanied by the release (–) of 414 kJ of energy (ΔH), referred to as the heat of reaction. From earlier considerations it is clear that electronic rearrangement (reaction or bond formation) takes place because the resulting solid product (NaCl) is at a lower energy state than the sum total of the energies of the original components.

The energetics associated with ionic bond formation may be determined quantitatively by considering the energy changes associated with the individual steps leading from the starting materials to the final product (Haber-Born cycle).

The bond formation in NaCl may be formally presented as an electron-transfer reaction:

$$\overset{1e-}{Na + Cl} \rightarrow Na^+ + Cl^-$$

The reactions involved in this process which result in the formation of 1 mole of solid salt are:

(1) Lonization of Na:

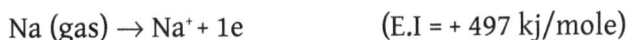

$$Na\ (gas) \rightarrow Na^+ + 1e \qquad (E.I = +\ 497\ kj/mole)$$

The energy change associated with this step, energy of ionization (E.I), is +497 kJ/mole.

(2) Acquisition of one electron by Cl:

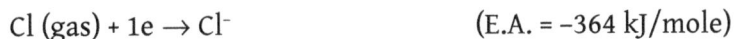

$$Cl\ (gas) + 1e \rightarrow Cl^- \qquad (E.A. = -364\ kJ/mole)$$

The energy change of one electron associated with this step, *electron affinity* (E.I.), is –364 kJ/mole. The minus sign reflects an energy release, a lowering of the energy state associated with the achievement of stable rare gas configuration by chlorine.

So far, the energy balance appears positive (ΔE = +133 kJ); this means the reaction is not favored since the final products are at a higher energy state than the starting products. However, there are additional steps involved since:

(3) Vaporization of Na:

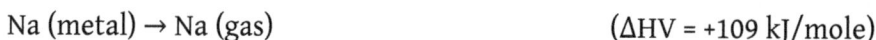

Na (metal) → Na (gas)   (ΔHV = +109 kJ/mole)

The energy required to transform Na (metal) into Na (gas), the latent heat of vaporization (ΔHV), is 109 kJ/mole (now reaction appears even less favorable).

(4) Dissociation of Cl_2:

Cl_2 → 2Cl (E.D. = +242 kJ)

The energy associated with breaking up the (stable) chlorine molecule into two reactive chlorine atoms, dissociation energy (E.D.), is +242 kJ/mole. (Since the formation of one mole NaCl involves only 1 mole Cl, and since 2 moles are formed from one mole Cl2, the energy required in this step is 1/2 E.D., or 121 kJ.)

The total energy change associated with reactions (1) through (4), ΔHT = +364 kJ This indicates that one or more basic reactions which lead to an overall decrease in energy are still unconsidered.

(5) Unconsidered as yet is the coulombic attraction of the reaction products which are of opposite charge and an energy term associated with the formation of the "solid state" product, NaCl. The two ionic species, originally at zero energy (infinite distance of separation), attract each other with an accompanying energy change (decrease).

The energetics associated with the approach of the reaction partners is best considered by fixing the position of one, for example Na^+, and letting the other (Cl^-) approach: as Cl^- approaches Na^+, E decreases according to Coulomb's law with $e^2/4\pi\varepsilon_o r$. With the approach of the two oppositely charged ions, the outermost electronic shells will come into contact and a repulsive force will become active as the shells interpenetrate. The repulsive force (Erep) increases with $+b/r^{12}$ and thus is only active in the immediate vicinity of the sodium ion, but at that stage increases rapidly. The two ions at close proximity are under the influence of both attractive and repulsive forces and will assume a distance of separation at which the two forces balance each other - a distance which is referred to as the equilibrium separation (r_o) and which corresponds to the energy minimum for the NaCl molecules:

$$E_{coul} = -\frac{e^2}{4\pi\varepsilon_o r_o} + \frac{b}{r_o^{12}}$$

where e = electronic charge, ε_o = permittivity of free space (8.85×10^{-12} F/m) and (b) is a constant. (r_o for gaseous molecules may be obtained from physical measurements.)

If the presently considered coulombic (attractive) energy term is taken into consideration as reaction (5), it will decrease the overall positive energy change, ΔHT; it will, however, not make it change sign from (+) to (–). Thus, NaCl molecules in gaseous form are not the final reaction product; nor would reaction occur if ΔHT remained positive.

Considering an ionically bonded, gaseous Na^+Cl^- molecule, it is clear that its electrostatic forces (+) and (–) are not saturated - they remain active in all possible directions. This means that Cl^- will attract Na^+ ions from other directions as will the Na^+ atoms attract additional Cl^- ions. The result of these attractive forces in all directions is the formation of a "giant size" ionic body - a "solid" body of macroscopic dimensions. From the preceding it is recognized that the minimum energy configuration is given by a body in which Na^+ and Cl^- are arranged with extreme periodicity and order since any ion located outside of its "equilibrium position" will be in a "higher energy configuration"; such an ordered body is referred to as a crystalline solid, or frequently just called a solid.

The total energy change associated with the formation of one mole of crystalline ionic solid from its ionic constituents is given as:

$$E_{cryst} = - \frac{M \, N_A \, e^2 (Q_1 Q_2)}{4\pi\varepsilon_0 r_0} \left(1 - \frac{1}{n}\right)$$

where M = 1.747 ("Madelung" constant for NaCl, reflecting multiple interactions for the particular geometric arrangement of ions in the solid), NA = Avogadro's number, Q = number of charges per ion (1 for Na^+ and Cl^-), ro= equilibrium distance of separation of ions, and n = repulsive exponent (n = 12 for NaCl).

The relationship for the crystal energy (ΔE_{cryst}) is readily obtained from equilibrium energy considerations:

$$E_{coul} = - \frac{e^2}{4\pi\varepsilon_0 r} + \frac{b}{4\pi\varepsilon_0 r^n}$$

At equilibrium distance of ion separation (r_0), (dE/dr) = 0. Thus:

$$\left(\frac{dE}{dr}\right)_{r_0} = \frac{e^2}{4\pi\varepsilon_0 r_0^2} - \frac{nb}{4\pi\varepsilon_0 r_0^{n+1}} = 0$$

and

$$b = \frac{e^2 r_0^{n-1}}{n}$$

and

$$E_{o(Coul)} = -\frac{e^2}{4\pi\varepsilon_o r_o} + \frac{e^2}{n\,4\pi\varepsilon_o r_o}$$

$$E_{o(Coul)} = -\frac{e^2}{4\pi\varepsilon_o r_o}\left(1 - \frac{1}{n}\right)$$

For a "molar" crystal with ionic charges Q_1 and Q_2, the molar $\Delta E_{o(Coul)}$ is thus given as:

$$\Delta E_{cryst} = -\frac{MN_A Q_1 Q_2 e^2}{4\pi\varepsilon_o r_o}\left(1 - \frac{1}{n}\right)$$

Here M and N_A stand, as indicated above, for the conventional terms, the Madelung constant and Avogadro's number.]

For the present system (NaCl):

$\Delta E_{cryst} = -777$ kJ/mole.

Now considering reactions (1) through (5):

$\Delta H_{Total} = -414$ kJ/mole

which is identical with the value experimentally determined and given for the reaction:

Na (metal) + Cl_2 (gas) $\rightarrow$ NaCl (solid)

$\Delta H_{Reaction} = -414$ kJ/mole

In the Haber-Born cycle, the reaction energy (ΔH) associated with the formation of NaCl from Na + Cl_2 may be summarized as:

$$\Delta H = E.I. + E.A. + \Delta H_V + 1/2\ E.D. + \Delta H_{cryst}$$

ΔH, the heat of reaction, may thus be obtained from the energetics of the steps leading to the end product. In most instances, however, the reaction energy (ΔH) is determined experimentally in a calorimeter and the Haber-Born cycle is used to obtain the value of ΔE_{cryst} or E.A., which are both extremely hard to come by.

Conclusions: A primary drive for atomic interactions leading to "bonding" is the achievement of valence shell octets which exhibit a high degree of stability. If atoms on the left and right side of the Periodic Table interact, i.e. atoms with a large difference in electron affinity (ΔE.A.), stabilization is achieved by electron octet formation through charge transfer. The reaction products exhibit opposite charges (cations, anions) and are subject to Coulombic attraction – ionic bonds are formed. Since the electrostatic forces are non-directional and non-saturated, energy minimization will result in the formation of macroscopic bodies that are highly ordered on the atomic scale, crystalline ionic solids. Ionic solids have mostly predictable, basic properties:

- ⊙ atomic arrangements are a function of the ion size ratio, the charge ratio of ions, and their electronic structure;

- ⊙ electrical and thermal conductivities are expected to be low because the high stability of the octets formed results in bound electrons which do not contribute to conduction (there is a large "energy gap" to be crossed for electrons to move into a higher energy state);

- ⊙ optically, ionic solids are mostly transparent, or translucent, reflecting octet stability of the electrons and macro- or micro-crystallinity;

- ⊙ melting points are high, increasing with the electronic charges on the cations and anions;

- ⊙ ionic solids are normally hard and brittle.

12.2.6 COVALENT BONDING

Wave Mechanical Concepts and Conclusions

In 1924 L. DeBroglie advanced the hypothesis that all matter in motion possesses wave properties and can be attributed a particle wavelength

$$\lambda P = h/mv$$

where h is the Planck constant, m is the mass of the matter and v is its velocity.

The credibility of this hypothesis was established by Davisson and Germer in 1927 when they demonstrated that electrons (like electromagnetic radiation) are diffracted by crystal lattices. An important consequence of the dual nature of matter (it exhibits both particle and wave properties) is the uncertainty principle established in 1927 by W. Heisenberg. It states that it is impossible to simultaneously know with certainty both the momentum and position of a moving particle:

$$(\Delta p_x)(\Delta x) \geq h$$

We can paraphrase the uncertainty principle in the following manner: If the energy of a particle is known (measured) with high precision, its location is associated with a high degree of uncertainty.

If electrons occupied simple orbits (as postulated by Bohr-Sommerfeld), their momentum and position could be determined exactly at any moment - in violation of the uncertainty principle. According to Heisenberg, if the energy of an electron is specified with precision (sharpness of spectral lines), its location can only be specified in terms of the probability of finding this electron in a certain location (volume element). These arguments give rise to the concepts of probability density and electron cloud which are inherent to the wave-mechanical electron concept emanating from the solutions of Schrödinger's wave equation which relates the energy of an orbiting electron to its wave properties. When solving exactly the Schrödinger wave equation for an electron in a hydrogen atom, a quantization results according to which electrons can only assume certain energies which are in quantitative agreement with those obtained from the Bohr theory.

In comparison to the Bohr theory, significant differences are observed for the orbital quantization (orbital quantum number l) which specifies the orbital shape. For n = 1, l = 0 (1s orbital), wave mechanics predicts a spherical electron density distribution with a maximum probability density (Ψ^2) at a distance of 0.529 Å (a_o) from the nucleus. For l = 1 (p orbitals), however, it is found that the orbitals (three) form lobes aligned with rectangular coordinates. For l = 2 (d orbitals), five complex orbital configurations are obtained.

Previous considerations suggest that all elements attempt to assume a stable octet configuration with eight electrons in the valence shell. (For hydrogen, with only one shell occupied, the stable configuration consists of two electrons in the K shell - which is the maximum number of electrons that can be accommodated.) Stable octet formation is possible through electron transfer (and ionic bond formation) when, for example, elements in columns IA, IIA and IIIA react with elements in columns VA, VIA and VIIA, respectively. In these instances, the reaction partners exhibit rather pronounced differences in electron affinity and upon reaction one assumes octet configuration by losing one or more electrons while the other does so by acquiring the missing number of electrons.

This mechanism is clearly not possible if H reacts with H to form an H2 molecule where two atoms are bonded together. The same argument holds for the formation of N2, Cl2 and O2 molecules. Inert gas configuration (octet configuration) in such elements is achieved by a mechanism called orbital sharing and the resulting bond is called covalent, or electron–pair bond.

A covalent bond is somewhat more difficult to visualize than an ionic or electrovalent bond because it involves the sharing of a pair of electrons between atoms. The stability of this bond can be attributed to the complex mutual attraction of two positively charged nuclei by the shared pair of electrons. In principle, the bond can be understood if it is recognized that both electrons in the bonding orbital spend more time between the two nuclei than around them and thus must exercise attractive forces which constitute the bond. In this arrangement it is clear that each electron, regardless of its source, exerts an attractive force on each of the "bonded" nuclei. The pair of electrons in a covalent bond is unique to the extent that the Pauli exclusion principle precludes the presence of additional electrons in the same orbital. Furthermore, the pairing phenomenon neutralizes the separate electronic spins of the single electrons, and the resulting electron pair with its zero spin momentum interacts less strongly with its surroundings than do two independent electrons.

LEWIS NOTATION: In LEWIS notation the covalent molecular bond is indicated as a BAR or as two DOTS (standing for the paired electrons)

H-H or H:H

:C̈l:C̈l: or |C̄l|C̄l| or Cl−Cl;

$$CH_4 = \quad H-\underset{\displaystyle H}{\overset{\displaystyle H}{\underset{|}{\overset{|}{C}}}}-H$$

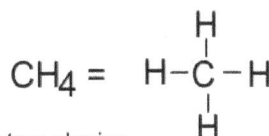

Formal valence shell octet stabilization can be achieved by electron sharing, whereby one electron from each reaction partner share - spin paired - the molecular bonding orbital.

Figure 3: Lewis notation

Quantum mechanics makes it possible to rigorously describe these bonds for very simple cases such as the hydrogen molecule, which is composed of two protons and two electrons. It can thus be shown that the potential energy for the system reaches a minimum for a certain equilibrium distance between the nuclei, with increased electron density between the nuclei. At shorter distances between the nuclei repulsive forces are found to increase very rapidly.

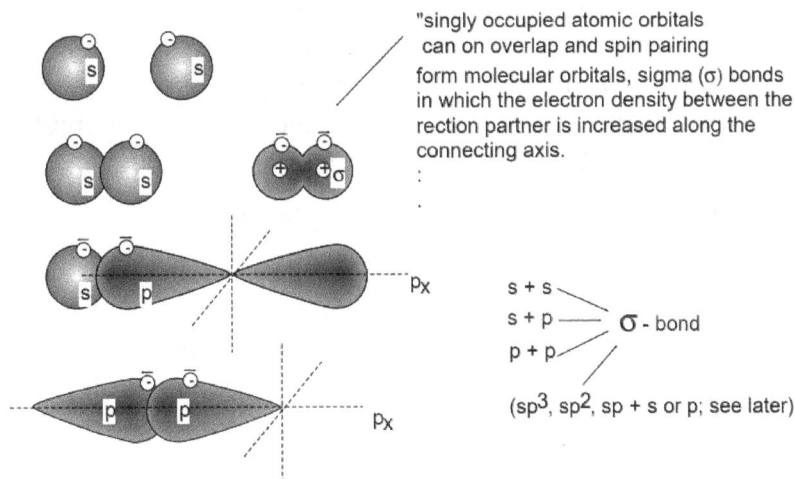

"singly occupied atomic orbitals can on overlap and spin pairing form molecular orbitals, sigma (σ) bonds in which the electron density between the rection partner is increased along the connecting axis.

s + s
s + p   σ - bond
p + p

(sp³, sp², sp + s or p; see later)

Figure 4: Covalent bond formation

In the hydrogen molecule (H2), the two hydrogen atoms are effectively linked together by one molecular electron orbital, termed a σ orbital, which comprises both atoms and contains two electrons. (fig. 4) Each of these two electrons can be considered to have originated from one of the two atoms - they were originally both 1s electrons with the same spin value (s = +1/2). In the molecular orbit comprising both atoms, the spins of the two electrons must align anti-parallel (opposite spin). This spin-pairing process results in a considerable release of energy and thus contributes significantly to the stability (strength) of the covalent bond formed.

It is interesting that according to Newtonian mechanics, no stable configuration can arise from the placement of two electrons into the same region (orbit) - wave mechanics , however, predicts increased stability from such configurations. (Similar spin-pairing occurs in the filling of atomic orbitals.) In the molecular case, spin-pairing has the consequence that both the probability distributions (Ψ^2) and spatial distribution of electrons are such that maximum overlap of orbitals of combining atoms occurs.

The bond formed in hydrogen is a single bond, containing two electrons with paired spins. This formulation predicts that there will be a fixed internuclear separation which, for the hydrogen molecule, has the value of 0.74 Å. [σ bonds are formed not only by s orbital overlap, but also by p-s and p-p orbital overlap as well as by the overlap of s or p orbitals with hybridized orbitals.

It is important to recognize that octet stabilization by electron orbital sharing results in bond properties which differ fundamentally from those encountered in ionic (electrovalent) bonding: with the formation of the covalent bond between the hydrogen atoms (H2 molecule formation) the bond forming capabilities of the two hydrogen atoms are saturated; the final product is a distinct molecule (H2) rather than a giant-sized solid body which is obtained as the final product with ionic bond formation. Covalent solid bodies, however, also do exist: they are formed if the elements involved have the capability of forming more than one bond. For example, carbon will form four covalent bonds in tetrahedral configuration - the result is diamond, a covalently bonded, three dimensional network.

12.2.7 Diatomic Molecules Involving Dissimilar Atoms

Consider the compound hydrogen chloride (in the gaseous, liquid or solid state). This compound is not ionic (because the energy state on complete ionization would be higher. Instead, bonding between hydrogen and chlorine atoms is accomplished by the sharing of electrons in a molecular orbital, thus forming a single covalent bond.

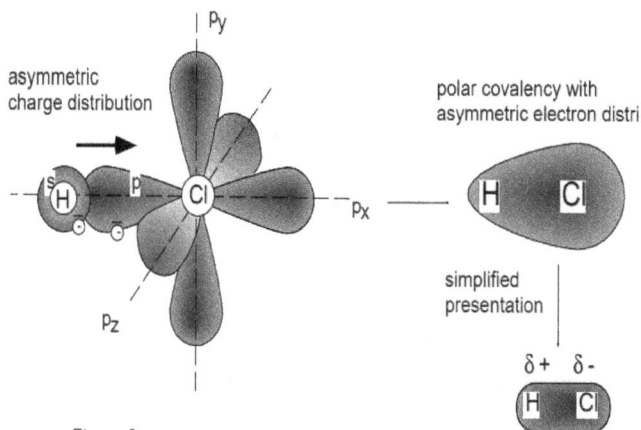

The electrons involved are the 1s electron of the hydrogen atom and the unpaired 3pz electron of the chlorine atom. After spin-pairing, a typical σ bond results in which each atom attains a noble gas structure: in HCl the hydrogen atom is "involved" with two electrons, as in helium, and the chlorine atom with 18 electrons, as in argon. The other (inner) electrons of chlorine do not participate in the bonding; they are termed non-bonding.

According to the above considerations, the bonding of HCl may be taken as very similar to that of H2. However, in the hydrogen molecule the electrons participating in the σ bond formation do not favor the proximity of either of the hydrogen nuclei - instead, they are equally shared. On the other hand, a preference for one nucleus (chlorine) is shown by the electrons of the σ bond in hydrogen chloride. This is because electrons, for energetic reasons, favor the environment of the more electronegative atom; in hydrogen chloride this is the chlorine atom and, consequently, the σ bond electrons spend more time in the vicinity of the chlorine atom. This situation results in the chlorine end of the molecule being fractionally negatively charged (δ^-) and the hydrogen end being fractionally positively charged (δ^+). We denote such an internal charge redistribution by the symbolism $H\delta^+ - Cl\delta^-$, where the δ sign indicates a partial electronic charge; the bond is said to be polar. (The hydrogen molecule does not a priori exhibit such an "asymmetric charge distribution"; it may, however, acquire a temporary polarization.) Molecules with asymmetric electronic charge distribution have a permanent dipole moment, the value of which is given by the product of the fractional charge (δ^+ must be equal to δ^-) and the distance of charge separation (L). Although dipole moments can be measured, their values do not allow us to directly calculate the polarity of bonds between atoms since the detailed internal charge distribution is unknown. The direction of electron drift and, to some extent, the magnitude of its effect can be estimated from the magnitude of the difference in electron affinity between the reaction partners. Because of a limited data base on electron affinity, L. Pauling introduced a related term, the relative electronegativity. From the above considerations it should be clear that no sharp dividing line exists between ionic and covalent bonding. We might consider a completely ionic bond to result in cases where the electron drift is such that one atom (the cation) becomes entirely deficient in one or more electrons, and the other atom (the anion) becomes correspondingly electron-rich, the bonding electrons being entirely under the influence of the latter. Hydrogen halides are σ bonded and have dipole moments; the bonds are referred to as polar covalency. The homonuclear diatomic molecules formed by the halogens, viz. F_2, Cl_2, Br_2 and I_2, are all σ-bonded systems involving spin-pairing of the various pz electrons (2pz, 3pz, 4pz and 5pz respectively). They have no permanent dipole moments.

12.2.8 Energetics of Covalent Bonding

Pauling extensively treated the energetics of polar covalencies, encountered in all heteronuclear systems (such as H–Cl) which have permanent dipole moments ($H\delta^+$ – $Cl\delta^-$). His approach visualizes the bonding to consist of two components - a pure covalency and an ionic bonding component with attraction resulting from the interaction of the fractional charges on the nuclei involved.

Pauling determines the basic covalent bonding component, for example in the formation of HCl, from the experimentally obtained bond energies associated with the molecular species of the components, H2 and Cl2. For this purpose he made the basic assumption that the covalent bond component between the dissimilar atoms is given by the geometric mean of the pure covalent bond energies associated with these molecular species. Thus:

Pure Covalent Bond Energy of HCl:

$$BE_{HCl} = \sqrt{BE_{H_2} \times BE_{Cl_2}}$$

$$H_2: \quad H - H \quad (\text{Bond Energy} = BE_{H_2} = 430 \text{ kJ/mole})$$

$$Cl_2: \quad Cl - Cl \quad (\text{Bond Energy} = BE_{Cl} = 238 \text{ kJ/mole})$$

$$BE_{HCl} = \sqrt{430 \times 238} = \boxed{320 \text{ kJ/mole}}$$

The ionic contribution (Δ) to the polar covalent bonding is then obtained from the difference between the experimentally determined bond energy (for HCl, for example), which obviously must contain both components, and the calculated pure covalent bond energy:

$$\Delta = \left[BE_{HCl} (\text{experimental})\right] - \left[BE_{HCl} (\text{theoretical} - \text{covalent})\right]$$

(The experimentally determined BE_{HCl} = 426 kJ/mole.)

$$\Delta = 426 - 320 = \boxed{106 \text{ kJ/mole}}$$

In connection with the presently discussed work, Linus Pauling established the now generally used electronegativity scale, or, better, the scale of relative electronegativities. This scale, included in the Periodic Table of the Elements, is extremely helpful since values of the electron affinity are known thus far for only very few elements.

The electronegativity (x) scale lists the relative tendency of the neutral elements to attract an additional electron. (The values of x are conventionally listed in electron Volts.) The scale listed was obtained by arbitrarily fixing the value of xH = 2.2. Pauling obtained the values for the other elements by relating differences in electronegativity of reaction partners to the fractional ionic character (ionic bonding component) of the bond established between them:

$$\text{experimental } BE_{AB} = \sqrt{BE_{AA} \times BE_{BB}} + k(x_A - x_B)^2$$

or

$$\Delta = 96.3 \ (x_A - x_B)^2 \ kJ$$

According to Pauling, the bonding character between two different elements may be defined as:

ionic bonding for: $\Delta x > 1.7$; covalent bonding for: $\Delta x < 1.7$

The fractional ionicity of polar covalent bonding as listed in the P/T is obtained by the relationship:

$$\% \text{ ionic bonding} = \left(1 - e^{-0.25(x_A - x_B)^2}\right) \times 100$$

12.2.9 Bonding In Polyatomic Molecules

In the formation of covalent bonds between atoms in polyatomic molecules, the conditions that atomic orbitals distort so that maximum overlap may be achieved when bonding occurs produces more extensive changes of orbital geometry than is the case with diatomic molecules. It will be remembered that the changes which result in a σ bond formation in diatomic molecules are a distortion of the s orbitals or p lobes so that strongest bond formation results from the maximum overlap between the two joined nuclei. For polyatomic molecules extensive alterations in the spatial disposition of atomic orbitals do occur, and very often the unique orbital geometry of the original atomic orbitals is completely lost . It is sometimes convenient to view these alterations as occurring in each atom prior to bonding by a process involving the mixing, or hybridization, of atomic orbitals.

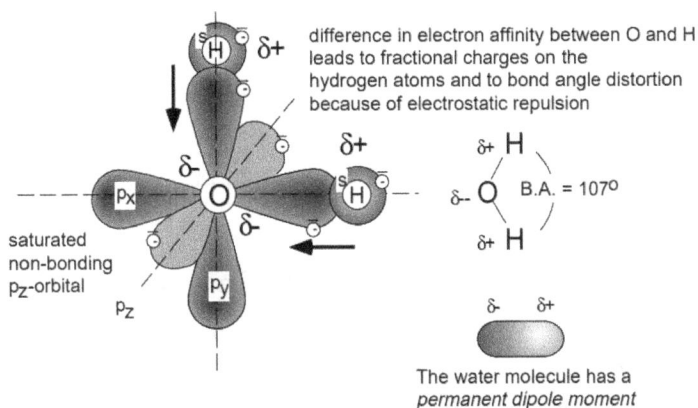

difference in electron affinity between O and H leads to fractional charges on the hydrogen atoms and to bond angle distortion because of electrostatic repulsion

B.A. = 107°

saturated non-bonding p_z-orbital

The water molecule has a *permanent dipole moment*

12.2.10 Bonding Involving Carbon

In methane (CH_4) the four outers, or valence, electrons of carbon are shared with the electrons of hydrogen; there is spin-pairing (resulting in bond formation) between each individual hydrogen electron and one of the carbon valence electrons. The noble gas structure is thus attained by each nucleus: the carbon nucleus "sees" eight outer electrons and each H nucleus "sees" two electrons. Consider now the orbitals which are involved in more detail. Each hydrogen has one spherical valence orbital (the 1s orbital) containing one electron, and covalent bond formation results from its distortion, overlap with the valence orbitals of carbon and spin-pairing.

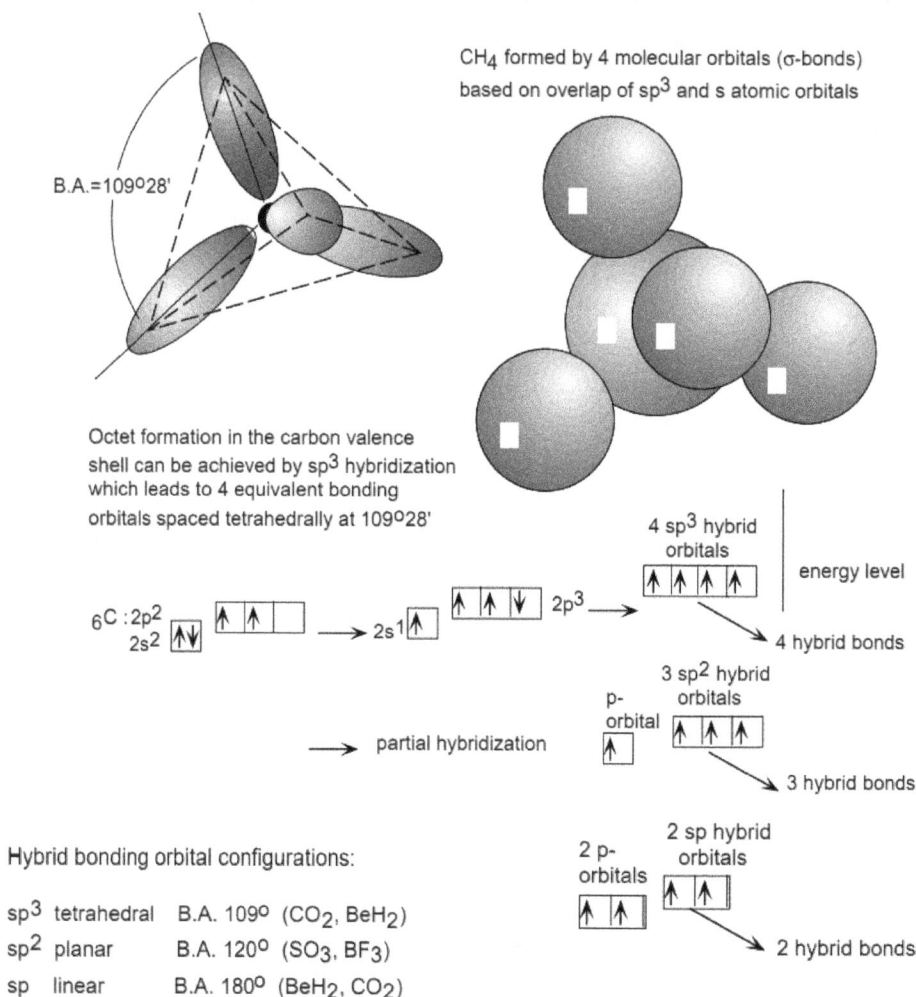

B.A.=109°28'

CH_4 formed by 4 molecular orbitals (σ-bonds) based on overlap of sp^3 and s atomic orbitals

Octet formation in the carbon valence shell can be achieved by sp^3 hybridization which leads to 4 equivalent bonding orbitals spaced tetrahedrally at 109°28'

$6C : 2p^2$
$2s^2$ → $2s^1$ → $2p^3$ →

4 sp^3 hybrid orbitals

energy level

4 hybrid bonds

partial hybridization

p-orbital

3 sp^2 hybrid orbitals

3 hybrid bonds

Hybrid bonding orbital configurations:

sp³ tetrahedral B.A. 109° (CO_2, BeH_2)
sp² planar B.A. 120° (SO_3, BF_3)
sp linear B.A. 180° (BeH_2, CO_2)

2 p-orbitals

2 sp hybrid orbitals

2 hybrid bonds

Figure 5: Orbital Hybridization

According to the Aufbau principle and Hund's rule, in its ground state carbon has two electrons in the filled K shell and four electrons in the L shell: two 2s electrons and two 2p electrons in singly occupied orbitals which are capable of covalent bond formation. This configuration provides, in principle, only two orbitals (2px,

2py) for covalent bond formation. However, with two covalencies carbon will not yield the desirable octet configuration. Such a configuration can be obtained if one of the two 2s electrons is "promoted" into the empty 2pz orbital since this process results in four singly occupied orbitals - all of which, being singly occupied, are capable of bond formation. The dissimilar singly occupied orbitals can assume (and therefore will assume) upon bond formation a lower energy configuration involving a process called hybridization whereby the four orbitals (of two different types) hybridize into four identical orbitals of maximum equal spacing from each other. Thus the hybridized orbitals (sp³ hybrids) are lobes emanating from the carbon atom into the corners of a tetrahedron, forming bond angles of 109°28'. sp³ hybridization is characteristic for carbon; however, other types of hybridization, sp² and sp, are encountered in other elements as well. Boron, for example, will tend to promote one of its two 2s electrons into a 2p state and, by hybridization, form three equivalent sp³ orbitals which assume planar orientation with band angles of 120°. Beryllium forms sp hybrid orbitals of linear orientation (the bond angle is 180°). All hybrid orbitals are capable of σ bond formation.

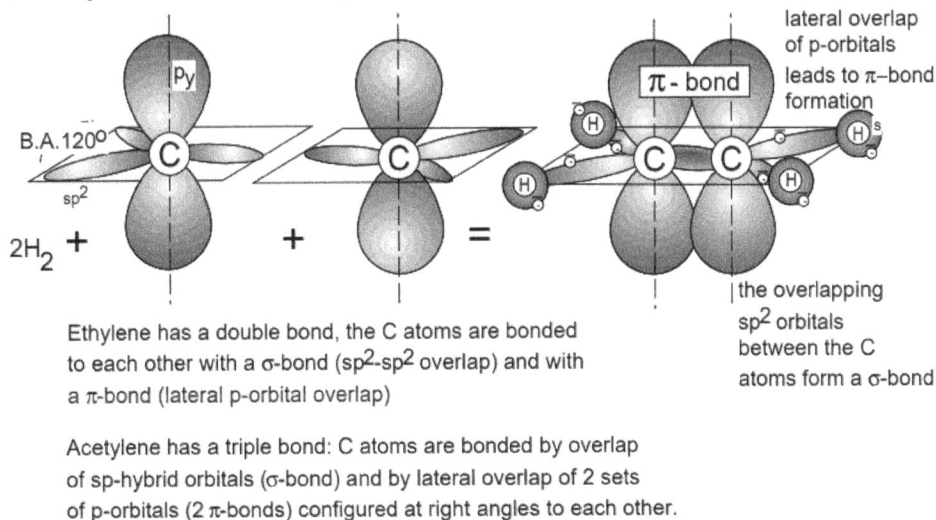

Ethylene has a double bond, the C atoms are bonded to each other with a σ-bond (sp2-sp2 overlap) and with a π-bond (lateral p-orbital overlap)

Acetylene has a triple bond: C atoms are bonded by overlap of sp-hybrid orbitals (σ-bond) and by lateral overlap of 2 sets of p-orbitals (2 π-bonds) configured at right angles to each other.

Figure 6: Double and Triple Bonds

The great diversity of carbon compounds (it forms more compounds than all the rest of the elements in the Periodic Table) can be attributed to the hybridization capability of the carbon orbitals (sp3, sp2 and sp). As a consequence, carbon forms not only axisymmetric σ bonds (through axial orbital overlap), but also π bonds (through lateral orbital overlap). In the compound ethane (H2C = CH2), the interactive carbon atoms undergo sp2 hybridization for a σ bond by overlap of two sp2 hybrid orbitals and form, in addition, a π bond by lateral overlap of the remaining non-hybridized p orbitals. (Double bonds involve one σ and one π bond.) In acetylene (HC = CH), we find sp hybridization and, by axial overlap, σ bond formation as well as lateral overlap of the remaining non-hybridized px and py orbitals which form two π bonds (fig.).

12.2.11 METALLIC BONDING

The stability of both covalent and metallic bonds may be regarded as arising from the potential energy lowering experienced by valence electrons under the influence of more than one nucleus. In metals, where the valence electrons are not as tightly bound to their ion cores, we cannot expect the formation of strong electron-pair bonds. The bond energies of known diatomic molecules of metallic elements are, in fact, smaller than those consisting of nonmetallic elements: 104 kJ/mole for Hg_2. Only diatomic molecules of the semi–metals have relatively high binding energies (385 kJ/mole for As2, 293 kJ/mole for Sb2 and 163 kJ/mole for Bi_2). These values reflect multiple bonding. Much greater stability is possible in larger aggregates of atoms such as bulk metals.

The known properties of metals, such as low electrical resistance and malleability, support the conceptual view that the valence electrons in metals never remain near any particular atom very long, but drift in a random manner through the lattice of ion cores. We may therefore visualize metals as a lattice of ion cores being held together by a gas of free electrons.

12.2. 12. SECONDARY (van der WAALS) BONDING

Primary bonding (ionic, covalent and metallic) is strong and the energies involved range from about 100 to 1000 kJ/mole. In contrast, secondary bonding is weak, involving energies ranging from about 0.1 to 10 kJ/mole. While this type of bonding, also referred to as "residual", is weak, it is essential in the functioning of our environment. Coke would likely be gaseous and not a bubbly, refreshing brew were it not for secondary bonding, nor would catalytic converters function. The energy difference between the liquid and vapor states of a given system is given by the heat of vaporization, i.e. the heat required to convert a given liquid into a vapor (normally) at the boiling point temperature at 1 atm pressure. The energy difference is due to intermolecular attraction between molecules at close distance of separation. This phenomenon of attraction through secondary bonding can best be considered between a single pair of molecules, but recognizing that the forces are of longer range. Four types of intermolecular forces can be identified.

1. Dipole-Dipole Interaction: Molecules with permanent dipoles (such as water, alcohol and other organic compounds with functional groups) exert a net attractive force on each other as a result of varying degrees of alignment of oppositely charged portions of the molecules (fig.) . For two polar molecules with a dipole moment of (μ) separated by a distance of (r), the energy of attraction can be quantified as:

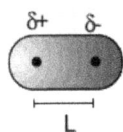

Schematic presentation of a dipole, exhibiting a Dipole-moment (μ) which is given by:

$$\mu = L.\partial q$$

where L is the distance of separation of the fractional electronic charges $\partial q+$ and $\partial q-$ (in abbreviated form given as $\delta+$ and $\delta-$). In molecules the dipole moment is given by the vector sum of the polar bonds. Thus, because of molecular geometry (symmetry), molecules such as CCl_4 exhibit no dipole-moment.

Dipole-Dipole Interaction:

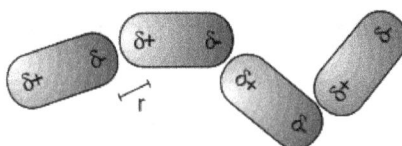

Figure 7

where μ is the dipole moment, r is the distance of approach of the oppositely charged molecular portions, k is the Boltzmann constant and T is the absolute temperature in K.

The molar energies of attraction associated with dipole-dipole interaction range from 0 to about 10 kJ/mole. These forces are primarily responsible for the liquid state (at room temperature) of most polar organic molecules; they are a contributing factor for H2O to be liquid at room temperature, and are responsible for alcohol being a liquid.

2. Dipole-Induced Dipole Interaction: A dipole in one molecule can interact with and polarize the electrons of a neighboring non-polar molecule, thus generating an induced dipole which will experience an attractive force with the polarizing, polar dipole. P. Debye showed that in a molecule with a "polarizability" of (α) the attractive potential arising from dipole-induced dipole interaction is given as:

$$E_{Dipole-Induced\,dipole} = -\frac{2\alpha\mu^2}{Dr^6}$$

Dipole-Induced Dipole Interaction:

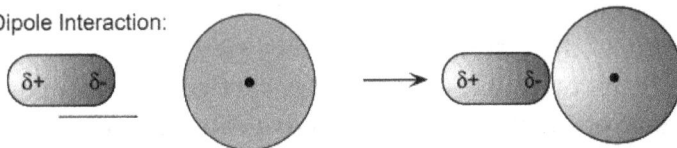

Figure 3: Induced dipole interaction is important in aqueous solutions and very effective during adsorption of inert molecules on active solid substrates.

3. London Dispersion Forces: It is a well-known fact that all substances, including rare gases and hydrogen, assume liquid state at finite temperature, an indication of the existence of attractive interatomic and intermolecular forces, even in the absence of permanent dipole systems. The origin of this force has been proposed by F. London in 1930. Accordingly, orbiting electrons will at any instance generate a "temporary dipole", the configuration of which changes as the electrons move. Since all atoms of a given system similarly experience temporary instantaneous dipole moments, their effect is expected to be cancelled because of the statistically random orientation of dipoles. It is evident that, should the dipoles be synchronized in a given assembly of atoms, then a net attractive force would result. But since such an attractive force constitutes a lowering of the energy of a given system, synchronization can and will take place because all systems will attempt to assume minimum energy configuration. The London dispersion force can be formulated as:

$$E_{London} = -K \frac{\alpha^2}{r^6}$$

The attractive London forces are small, as manifested by the very low boiling points of the smaller rare gases, of hydrogen and nitrogen.

4. Hydrogen Bonding: The short and long range dipole interactions calculated from molecular dipole moments are inadequate in explaining a multitude of phenomena in organic as well as some inorganic systems. L. Pauling studied such London Dispersion (Fluctuating Dipole Interaction) Electron motion in atoms generates fluctuating dipoles. Since the polarization is random attractions are cancelled by repulsions.

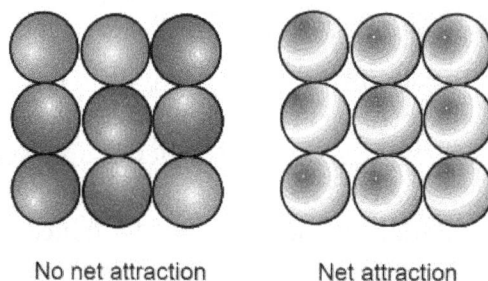

No net attraction   Net attraction

Figure 9

London: Upon synchronization of electron motion inter atomic attraction will be established in an assembly of atoms. Since this attraction constitutes a lowering of the energy of the system - attraction by synchronization will take place - as all systems attempt to assume lowest energy configuration phenomena and concluded the existence of highly specific attractive interaction between hydrogen that is acidic (carries a fractional positive charge) and the elements O, F, N and, to a lesser extent, S in both organic and inorganic molecules. This interaction, which can as yet not be formulated, is referred to as hydrogen bonding; its magnitude, ranging up to 40 kJ/mole, is significantly larger than that of any other secondary bonding type. Hydrogen bonding is considered instrumental in controlling most properties of water, is a key element in the structure of nucleic acid and thought to be an essential component in memory functions of the human brain.

13 | CHALLENGES AND FUTURE DIRECTIONS

13.1 INTRODUCTION: AN OVERVIEW

The realm of chemical bonding stands as one of the fundamental pillars of chemistry, governing the interactions between atoms and molecules that underpin the entirety of the material world. From the simplest diatomic compounds to the complex molecular structures essential for life, understanding the nature of chemical bonds has been central to unraveling the mysteries of matter and driving scientific innovation. Over the decades, intensive research and theoretical developments have unveiled the intricacies of bonding phenomena, leading to profound insights into molecular structure, reactivity, and material properties.

Yet, the study of chemical bonding continues to present formidable challenges and beckons toward intriguing future directions. At its core, the concept of chemical bonding grapples with the delicate balance of forces that bind atoms together, encompassing a diverse array of interactions spanning from covalent bonds, where electrons are shared between atoms, to ionic bonds, where electrons are transferred outright. Moreover, the emergence of concepts such as hydrogen bonding, metal coordination, and aromaticity has further enriched our understanding of bonding diversity and complexity.

However, as our comprehension of chemical bonding deepens, so too does our recognition of its nuanced intricacies and unresolved puzzles. Challenges persist in accurately modeling and predicting the bonding behavior of increasingly complex molecules, particularly in systems where multiple types of interactions coexist. The interplay between electronic structure, molecular geometry, and environmental factors continues to elude complete characterization, demanding innovative computational methodologies and experimental techniques.

Furthermore, the exploration of chemical bonding extends beyond traditional organic and inorganic realms into interdisciplinary frontiers such as materials science, catalysis, and nanotechnology. In these domains, tailored manipulation of bonding interactions holds the key to unlocking revolutionary advances in energy storage, drug design, and environmental remediation. The quest for novel materials with enhanced functionalities and unprecedented properties propels the pursuit of unconventional bonding motifs and exotic molecular architectures.

Looking ahead, the future of chemical bonding research is poised to embark on a trajectory marked by interdisciplinary collaboration, computational sophistication, and experimental ingenuity. The integration of quantum mechanics, machine learning, and high-throughput experimentation promises to unravel hitherto inaccessible realms of chemical space and unveil hidden patterns in bonding landscapes. Moreover, the advent of cutting-edge spectroscopic and imaging techniques offers unprecedented glimpses into the dynamic evolution of chemical bonds under diverse conditions, paving the way for real-time monitoring and control of molecular transformations.

In essence, the challenges and future directions of chemical bonding epitomize the enduring quest of scientific inquiry—to unravel the mysteries of nature, harness the power of molecular interactions, and forge new frontiers of knowledge and discovery. As we stand at the precipice of a new era in chemistry, the journey toward elucidating the secrets of chemical bonding beckons with boundless promise and untold revelations.

13.2 CURRENT CHALLENGES IN UNDERSTANDING CHEMICAL BONDING

Understanding chemical bonding remains a central challenge in contemporary chemistry, despite significant advancements in theory, computation, and experimental techniques. Several pressing challenges persist in unraveling the intricacies of chemical bonding:

13.2.1 Complexity of Molecular Systems: As chemists delve into increasingly complex molecular systems, the understanding of chemical bonding becomes more intricate. Molecular structures can exhibit diverse bonding motifs, including covalent, ionic, metallic, and weak interactions such as hydrogen bonding and van der Waals forces. Deciphering the interplay between these different bonding types in complex molecules remains a formidable challenge.

13.2.2 Quantum Mechanical Effects: Chemical bonding is fundamentally governed by quantum mechanics, which describes the behavior of electrons within atoms and molecules. However, accurately modeling the electronic structure of molecules and predicting their bonding properties computationally is still challenging, particularly for large molecules or systems with strong electron correlation effects.

13.2.3 Non-Covalent Interactions: Non-covalent interactions, such as hydrogen bonding, π-π stacking, and ion-dipole interactions, play crucial roles in molecular recognition, supramolecular chemistry, and biomolecular interactions. Understanding the subtle balance between these interactions and their influence on molecular structure, reactivity, and properties remains a complex puzzle.

13.2.4 Dynamic Nature of Bonding: Chemical bonds are not static entities but can undergo dynamic changes in response to external stimuli or environmental conditions. Investigating the dynamics of bond formation and cleavage, as well as the role of solvent effects and temperature, presents challenges for experimental observation and theoretical modeling.

13.2.5 Interface of Theory and Experiment: Bridging the gap between theoretical predictions and experimental observations is a perennial challenge in understanding chemical bonding. While theoretical models provide valuable insights into bonding mechanisms, experimental validation is essential for confirming theoretical predictions and refining our understanding of chemical interactions.

13.2.6 Emerging Materials and Nanostructures: The design and characterization of novel materials, nanomaterials, and nanostructures pose unique challenges in understanding chemical bonding at the nanoscale. Manipulating bonding interactions to engineer materials with specific properties requires a deep understanding of molecular structure-property relationships and the ability to control bonding at the atomic level.

13.2.7 Interdisciplinary Nature of Bonding: Chemical bonding is inherently interdisciplinary, spanning fields such as chemistry, physics, materials science, and biology. Collaborative efforts across disciplines are essential for addressing complex bonding phenomena and developing innovative approaches to tackle longstanding challenges in chemical bonding research.

Addressing these challenges requires a concerted effort involving theoretical and computational chemistry, experimental spectroscopy and imaging techniques, as well as interdisciplinary collaborations that transcend traditional boundaries. By confronting these challenges head-on, scientists can deepen our understanding of chemical bonding and pave the way for transformative advances in fields ranging from drug discovery and materials science to catalysis and nanotechnology.

13.3 EMERGING TRENDS AND TECHNOLOGIES

Emerging trends and technologies in chemical bonding are reshaping the landscape of modern chemistry, pushing the boundaries of our understanding and opening new avenues for exploration and innovation. At the forefront of this evolution is the integration of computational methods and machine learning algorithms, which have revolutionized our ability to predict, analyze, and manipulate chemical bonds with unprecedented accuracy and efficiency. Advanced quantum mechanical techniques, such as density functional theory (DFT) and ab initio molecular dynamics, enable the simulation and characterization of complex bonding phenomena across diverse chemical systems, providing insights into molecular structure, energetics, and reactivity. Complementing these approaches, machine learning algorithms harness vast datasets to extract patterns, predict bonding behaviors, and guide the design of novel materials with tailored properties, accelerating the pace of materials discovery and optimization.

High-resolution spectroscopic and imaging techniques represent another frontier in chemical bonding research, offering unprecedented insights into molecular structure and dynamics at the atomic level. Techniques such as X-ray crystallography, nuclear magnetic resonance (NMR) spectroscopy, and scanning probe microscopy enable researchers to probe bonding interactions, conformational changes, and intermolecular forces with unparalleled precision and resolution, facilitating the rational design of functional materials and molecular devices. Furthermore, single-molecule and single-atom manipulation techniques provide the ability to probe and manipulate individual chemical bonds and molecular assemblies with atomic precision, paving the way for bottom-up construction of molecular architectures and nanoscale devices with tailored functionalities.

Multiscale modeling and simulation approaches have emerged as powerful tools for investigating chemical bonding phenomena across multiple length and time scales, capturing the interplay between electronic structure, molecular dynamics, and environmental effects. By integrating techniques from quantum mechanics to classical mechanics, these methods facilitate the study of complex systems such as biomolecules, polymers, and materials interfaces, enabling researchers to uncover fundamental principles governing bonding interactions and devise strategies for controlling and optimizing material properties. Moreover, interdisciplinary collaborations and cross-fertilization of ideas between chemistry, physics, materials science, and engineering drive innovation and synergistic approaches to address complex bonding challenges, fostering a dynamic and collaborative research ecosystem at the intersection of diverse disciplines.

Quantum computing represents a transformative frontier in chemical bonding research, offering the potential to simulate and optimize complex quantum mechanical systems with unprecedented speed and accuracy. Quantum simulators and algorithms enable researchers to explore electronic structure, molecular dynamics, and chemical reactivity in ways that are inaccessible to classical computers, unlocking new insights into bonding phenomena and paving the way for the design of novel materials and molecular devices with advanced functionalities. As technology continues to advance and interdisciplinary collaborations flourish, the exploration of chemical bonding promises to remain at the forefront of scientific inquiry, driving transformative discoveries and shaping the future of chemistry in profound and unexpected ways.

Chemical bonding, a cornerstone of chemistry, continues to evolve with emerging trends and technologies, shaping research directions and applications across various domains. Several key trends and technologies are driving the forefront of chemical bonding research:

13.3.1 Computational Chemistry Advances: Computational methods have revolutionized our understanding of chemical bonding by enabling the accurate prediction and analysis of molecular structures, energetics, and properties. Advanced quantum mechanical techniques, such as density functional theory (DFT), coupled cluster theory, and ab initio molecular dynamics, are increasingly applied to unravel complex bonding phenomena in diverse chemical systems.

13.3.2 Machine Learning and Data-Driven Approaches: Machine learning algorithms and data-driven approaches are revolutionizing chemical bonding research by leveraging vast datasets to extract patterns, predict bonding behaviors, and accelerate materials discovery. These methods complement traditional theoretical and experimental approaches, offering insights into complex bonding landscapes and guiding the design of novel materials with tailored properties.

13.3.3 High-Resolution Spectroscopy and Imaging Techniques: High-resolution spectroscopic and imaging techniques, such as X-ray crystallography, nuclear magnetic resonance (NMR) spectroscopy, and scanning probe microscopy, provide unprecedented insights into molecular structure and bonding interactions at the atomic level. Advancements in instrumentation and data analysis techniques enable researchers to probe bonding dynamics, conformational changes, and intermolecular interactions with unparalleled precision and resolution.

13.3.4 Single-Molecule and Single-Atom Manipulation: Single-molecule and single-atom manipulation techniques allow researchers to probe and manipulate individual chemical bonds and molecular assemblies with atomic precision. Techniques such as scanning tunneling microscopy (STM) and atomic force microscopy (AFM) enable the direct visualization and manipulation of chemical bonds, paving the way for bottom-up construction of molecular architectures and functional nanomaterials.

13.3.5 Multiscale Modeling and Simulation: Multiscale modeling and simulation approaches integrate techniques spanning from quantum mechanics to classical mechanics, enabling the investigation of bonding phenomena across multiple length and time scales. These methods facilitate the study of complex systems, such as biomolecules, polymers, and materials interfaces, by capturing the interplay between electronic structure, molecular dynamics, and environmental effects.

13.3.6 Interdisciplinary Collaborations and Cross-Fertilization: Chemical bonding research increasingly benefits from interdisciplinary collaborations that bridge traditional boundaries between chemistry, physics, materials science, biology, and engineering. Cross-fertilization of ideas, techniques, and methodologies accelerates innovation and fosters synergistic approaches to address complex bonding challenges and explore new frontiers of scientific inquiry.

13.3.7 Quantum Computing and Simulation: Quantum computing holds the potential to revolutionize chemical bonding research by enabling the simulation of complex quantum mechanical systems with unprecedented speed and accuracy. Quantum simulators and algorithms offer insights into electronic structure, molecular dynamics, and chemical reactivity that are inaccessible to classical computers, opening new avenues for understanding and manipulating chemical bonds.

By embracing these emerging trends and technologies, researchers are poised to unlock new dimensions of chemical bonding, unraveling the mysteries of molecular interactions, and harnessing the power of chemical bonds to drive innovation in fields ranging from materials science and catalysis to drug discovery and nanotechnology. As technology continues to advance, the exploration of chemical bonding promises to remain at the forefront of scientific inquiry, driving transformative discoveries and shaping the future of chemistry.

13.4 POTENTIAL AREAS FOR FUTURE RESEARCH

Future research in chemical bonding holds immense potential across diverse scientific and technological domains. One promising avenue is the advancement of computational methodologies, including the development of high-level quantum mechanical techniques and machine learning algorithms, to enable accurate prediction and analysis of bonding behaviours in complex systems. Understanding the intricacies of non-covalent interactions, such as hydrogen bonding and van der Waals forces, presents another frontier, with implications for drug design, materials science, and supramolecular chemistry. Additionally, exploring the dynamic nature of chemical bonds under varying conditions and stimuli, along with the rational design of materials and molecular devices based on bonding principles, offers exciting opportunities for innovation in nanotechnology, electronics, and catalysis. Furthermore, interdisciplinary collaborations aimed at integrating insights from chemistry, physics, biology, and engineering will be crucial for addressing complex societal challenges and driving transformative discoveries in chemical bonding research, paving the way for novel applications in fields ranging from biomedicine to sustainable materials and beyond. Top of Form

Future research in chemical bonding holds promise across a spectrum of interdisciplinary endeavors, driven by emerging technologies and scientific inquiries. Several potential areas for future exploration include:

13.4.1 Advanced Computational Methods: Further development and refinement of computational techniques, such as high-level quantum mechanical methods and machine learning algorithms, will enhance our ability to predict and understand chemical bonding in increasingly complex systems. Integration of quantum computing and simulation offers exciting prospects for exploring bonding phenomena with unprecedented accuracy and efficiency.

13.4.2 Non-Covalent Interactions: Investigating the role of non-covalent interactions, such as hydrogen bonding, π-π stacking, and van der Waals forces, in molecular recognition, supramolecular chemistry, and biomolecular interactions represents a fertile area for future research. Understanding the subtleties of these interactions and their impact on molecular structure and reactivity holds implications for drug design, materials science, and beyond.

13.4.3 Dynamic Bonding Processes: Exploring the dynamic nature of chemical bonding, including bond formation, cleavage, and reactivity, under varying environmental conditions and stimuli is an intriguing avenue for future investigation. Understanding the kinetics and thermodynamics of dynamic bonding processes is essential for designing functional materials and controlling molecular transformations.

13.4.4 Molecular Engineering and Design: Leveraging insights from chemical bonding research to engineer novel materials, catalysts, and molecular devices with tailored properties and functionalities is a promising direction for future exploration. Rational design principles informed by fundamental understanding of bonding interactions enable the development of materials with enhanced performance and versatility.

13.4.5 Nanoscale and Molecular Electronics: Investigating chemical bonding at the nanoscale and molecular level is essential for advancing molecular electronics, nanoelectronics, and quantum computing technologies. Understanding the electronic structure and transport properties of molecular junctions, interfaces, and nanostructures is critical for realizing next-generation electronic devices and quantum information processing platforms.

13.4.5 Biochemical and Biomedical Applications: Exploring the role of chemical bonding in biological systems, including enzyme catalysis, protein-ligand interactions, and drug-receptor binding, holds promise for advancing drug discovery, personalized medicine, and biotechnology. Understanding the molecular basis of disease and designing targeted therapeutics requires a deep understanding of bonding interactions in biological molecules.

13.4.6 Green Chemistry and Sustainable Materials: Developing environmentally sustainable processes and materials through innovative bonding strategies, catalysis, and molecular design represents a pressing challenge for future research. Harnessing renewable resources, minimizing waste, and reducing the environmental impact of chemical synthesis require novel approaches informed by principles of green chemistry and sustainable design.

13.4.7 Interdisciplinary Collaborations and Cross-Cutting Research: Promoting interdisciplinary collaborations and cross-cutting research initiatives that bridge traditional boundaries between chemistry, physics, materials science, biology, and engineering is essential for addressing complex societal challenges and driving transformative innovation in chemical bonding research.

By exploring these potential areas for future research, scientists can deepen our understanding of chemical bonding, unlock new frontiers of scientific inquiry, and harness the power of molecular interactions to address pressing societal needs and shape the future of chemistry in profound and impactful ways.

GLOSSARY

Anion: A negatively charged ion.

Bond: That which holds together atoms in molecules and ions in lattices.

Cation: A positively charged ion.

Coulomb's Law: A mathematical formula whose consequence is that negatively and positively charged particles attract each other and similarly charged species repel each other.

Covalent Bond: A bond that results from a sharing of electrons between nuclei.

Double bond: a covalent bond in which two pairs of electrons are shared by two atoms

Electro negativity: a measure of the ability of an atom in a molecule to draw bonding electrons to itself

Ion: an electrically charged particle obtained from an atom or a chemically bonded group of atoms by addin- or removing zn electrons

Ionic Bond: A bond that results from electrostatic attraction between oppositely charged ions. The cation is positively charged, while the anion is negatively charged.

Lattice: A regularly repeating three-dimensional array of atoms, molecules, or ions.

Lewis electron-dot system: a formula using dots to represent valence electrons

Molecular Orbital Theory: A description of bonding that combines atomic orbitals from each bonded atom to produce a set of molecular orbitals.

Molecular Orbital: A combination of atomic orbitals in molecular orbital theory that provides an orbital description of a molecule analogous to the atomic orbital description of atoms.

Molecule: A chemical species containing a covalent bond.

Octet rule: the tendency of atoms in molecules to have eight electrons in their valence shells (two for hydrogen atoms).

Polar covalent bond: a covalent bond in which the bonding electrons spend more time near one atom than near the other

Resonance: a representation in which we describe the electron structure of a molecule, having delocalized bonding by, writing all possible electron-dot formulas

Triple bond: a covalent bond in which three pairs of electrons are shared by two atoms

Valence electron: an electron in an atom outside the noble gas or pseudonoble-gas core

Valence Shell Electron Pair Repulsion Theory: A theory used to predict bonding geometries that states that electron pairs will be distributed about the central atom to minimize electron pair repulsions.

Metallic bond: An attraction between a positive metal ion and the electrons surrounding it.

π bonding orbital: molecular orbital formed by side-by-side overlap of atomic orbitals, in which the electron density is found on opposite sides of the internuclear axis

σ bonding orbital: molecular orbital in which the electron density is found along the axis of the bond

π* bonding orbital: antibonding molecular orbital formed by out of phase side-by-side overlap of atomic orbitals, in which the electron density is found on both sides of the internuclear axis, and there is a node between the nuclei

σ* bonding orbital: antibonding molecular orbital formed by out-of-phase overlap of atomic orbital along the axis of the bond, generating a node between the nuclei

acid anhydride: compound that reacts with water to form an acid or acidic solution

acid ionization: reaction involving the transfer of a proton from an acid to water, yielding hydronium ions and the conjugate base of the acid

acid ionization constant (Ka): equilibrium constant for an acid ionization reaction

acid-base indicator: weak acid or base whose conjugate partner imparts a different solution color; used in visual assessments of solution pH

acidic: a solution in which $[H3O+] > [OH-]$

actinide: inner transition metal in the bottom of the bottom two rows of the periodic table

actinide series: (also, actinoid series) actinium and the elements in the second row or the f-block, atomic numbers 89–103

active electrode: electrode that participates as a reactant or product in the oxidation-reduction reaction of an electrochemical cell; the mass of an active electrode changes during the oxidation-reduction reaction

alkali metal: element in group 1

alkaline battery: primary battery similar to a dry cell that uses an alkaline (often potassium hydroxide) electrolyte; designed to be an improved replacement for the dry cell, but with more energy storage and less electrolyte leakage than typical dry cell

alkaline earth metal: element in group 2

allotropes: two or more forms of the same element, in the same physical state, with different chemical structures

amorphous: solid material such as a glass that does not have a regular repeating component to its three-dimensional structure; a solid but not a crystal

amphiprotic: species that may either donate or accept a proton in a Bronsted-Lowry acid-base reaction

amphoteric: species that can act as either an acid or a base

anode: electrode in an electrochemical cell at which oxidation occurs

antibonding orbital: molecular orbital located outside of the region between two nuclei; electrons in an antibonding orbital destabilize the molecule

aqueous solution: solution for which water is the solvent

Aufbau principle: procedure in which the electron configuration of the elements is determined by "building" them in order of atomic numbers, adding one proton to the nucleus and one electron to the proper subshell at a time

autoionization: reaction between identical species yielding ionic products; for water, this reaction involves transfer of protons to yield hydronium and hydroxide ions

axial position: location in a trigonal bipyramidal geometry in which there is another atom at a 180° angle and the equatorial positions are at a 90° angle

base anhydride: metal oxide that behaves as a base towards acids

base ionization: reaction involving the transfer of a proton from water to a base, yielding hydroxide ions and the conjugate acid of the base

base ionization constant (Kb): equilibrium constant for a base ionization reaction

basic: a solution in which $[H3O+] < [OH-]$

battery: single or series of galvanic cells designed for use as a source of electrical power

bicarbonate anion: salt of the hydrogen carbonate ion HCO3-

bidentate ligand: ligand that coordinates to one central metal through coordinate bonds from two different atoms

binary acid: compound that contains hydrogen and one other element, bonded in a way that imparts acidic properties to the compound (ability to release H+ ions when dissolved in water)

binary compound: compound containing two different elements.

bismuth: heaviest member of group 15; a less reactive metal than other representative metals

bond angle: angle between any two covalent bonds that share a common atom

bond dipole moment: separation of charge in a bond that depends on the difference in electronegativity and the bond distance represented by partial charges or a vector

bond distance: (also, bond length) distance between the nuclei of two bonded atoms

bond energy: (also, bond dissociation energy) energy required to break a covalent bond in a gaseous substance

bond length: distance between the nuclei of two bonded atoms at which the lowest potential energy is achieved

bond order: number of pairs of electrons between two atoms; it can be found by the number of bonds in a Lewis structure or by the difference between the number of bonding and antibonding electrons divided by two

bonding orbital: molecular orbital located between two nuclei; electrons in a bonding orbital stabilize a molecule

borate: compound containing boron-oxygen bonds, typically with clusters or chains as a part of the chemical structure

Born-Haber cycle: thermochemical cycle relating the various energetic steps involved in the formation of an ionic solid from the relevant elements

Brønsted-Lowry acid: proton donor

Brønsted-Lowry base: proton acceptor

buffer: mixture of appreciable amounts of a weak acid-base pair the pH of a buffer resists change when small amounts of acid or base are added

buffer capacity: amount of an acid or base that can be added to a volume of a buffer solution before its pH changes significantly (usually by one pH unit)

carbonate: salt of the anion CO_3^{2-} often formed by the reaction of carbon dioxide with bases

cathode: electrode in an electrochemical cell at which reduction occurs

cathodic protection: approach to preventing corrosion of a metal object by connecting it to a sacrificial anode composed of a more readily oxidized metal

cell notation (schematic): symbolic representation of the components and reactions in an electrochemical cell

cell potential: difference in potential of the cathode and anode half-cells

central metal: ion or atom to which one or more ligands is attached through coordinate covalent bonds

chalcogen: element in group 16

chelate: complex formed from a polydentate ligand attached to a central metal

chelating ligand: ligand that attaches to a central metal ion by bonds from two or more donor atoms

chemical reduction: method of preparing a representative metal using a reducing agent

chemistry: study of the composition, properties, and interactions of matter

chlor-alkali process: electrolysis process for the synthesis of chlorine and sodium hydroxide

cis configuration: configuration of a geometrical isomer in which two similar groups are on the same side of an imaginary reference line on the molecule

color-change interval: range in pH over which the color change of an indicator is observed

common ion effect: effect on equilibrium when a substance with an ion in common with the dissolved species is added to the solution; causes a decrease in the solubility of an ionic species, or a decrease in the ionization of a weak acid or base

complex ion: ion consisting of a central atom surrounding molecules or ions called ligands via coordinate covalent bonds

concentrated: qualitative term for a solution containing solute at a relatively high concentration

concentration: quantitative measure of the relative amounts of solute and solvent present in a solution

concentration cell: galvanic cell comprising half-cells of identical composition but for the concentration of one redox reactant or product

conjugate acid: substance formed when a base gains a proton

conjugate base: substance formed when an acid loses a proton

coordinate covalent bond: (also, dative bond) covalent bond in which both electrons originated from the same atom

coordination compound: stable compound in which the central metal atom or ion acts as a Lewis acid and accepts one or more pairs of electrons

coordination number: number of coordinate covalent bonds to the central metal atom in a complex or the number of closest contacts to an atom in a crystalline form

coordination sphere: central metal atom or ion plus the attached ligands of a complex

core electron: electron in an atom that occupies the orbitals of the inner shells

corrosion: degradation of metal via a natural electrochemical process

coupled equilibria: system characterized the simultaneous establishment of two or more equilibrium reactions sharing one or more reactant or product

covalent bond: attractive force between the nuclei of a molecule's atoms and pairs of electrons between the atoms

covalent radius: one-half the distance between the nuclei of two identical atoms when they are joined by a covalent bond

crystal field splitting: difference in energy between the t2g and eg sets or t and e sets of orbitals

crystal field theory: model that explains the energies of the orbitals in transition metals in terms of electrostatic interactions with the ligands but does not include metal ligand bonding

d-block element: one of the elements coordination compound groups 3–11 with valence electrons in d orbitals

degenerate orbitals: orbitals that have the same energy

diamagnetism: phenomenon in which a material is not magnetic itself but is repelled by a magnetic field; it occurs when there are only paired electrons present

dilute: qualitative term for a solution containing solute at a relatively low concentration

dilution: process of adding solvent to a solution in order to lower the concentration of solutes

dipole moment: property of a molecule that describes the separation of charge determined by the sum of the individual bond moments based on the molecular structure

diprotic acid: acid containing two ionizable hydrogen atoms per molecule

diprotic base: base capable of accepting two protons

disproportionation reaction: chemical reaction where a single reactant is simultaneously reduced and oxidized; it is both the reducing agent and the oxidizing agent

dissociation constant (Kd): equilibrium constant for the decomposition of a complex ion into its components

dissolved: describes the process by which solute components are dispersed in a solvent

donor atom: atom in a ligand with a lone pair of electrons that forms a coordinate covalent bond to a central metal

Downs cell: electrochemical cell used for the commercial preparation of metallic sodium (and chlorine) from molten sodium chloride

dry cell: primary battery, also called a zinc-carbon battery, based on the spontaneous oxidation of zinc by manganese(IV)

e_g orbitals: set of two d orbitals that are oriented on the Cartesian axes for coordination complexes; in octahedral complexes, they are higher in energy than the t2g orbitals

effective nuclear charge: charge that leads to the Coulomb force exerted by the nucleus on an electron, calculated as the nuclear charge minus shielding

electrode potential: the potential of a cell in which the half-cell of interest acts as a cathode when connected to the standard hydrogen electrode

electrolysis: process using electrical energy to cause a nonspontaneous process to occur

electrolytic cell: electrochemical cell in which an external source of electrical electrolysis power is used to drive an otherwise nonspontaneous process

electron affinity: energy change associated with addition of an electron to a gaseous atom or ion

electron configuration: listing that identifies the electron occupancy of an atom's shells and subshells

electron-pair geometry: arrangement around a central atom of all regions of electron density (bonds, lone pairs, or unpaired electrons)

electronegativity: tendency of an atom to attract electrons in a bond to itself

entropy (S): state function that is a measure of the matter and/or energy dispersal within a system, determined by the number of system microstates; often described as a measure of the disorder of the system

equatorial position: one of the three positions in a trigonal bipyramidal geometry with 120° angles between them; the axial positions are located at a 90° angle

equilibrium: state of a reversible reaction in which the forward and reverse processes occur at equal rates

equilibrium constant (K): value of the reaction quotient for a system at equilibrium; may be expressed using concentrations (Kc) or partial pressures (Kp)

f-block element: (also, inner transition element) one of the elements with atomic numbers 58–71 or 90–103 that have valence electrons in f orbitals; they are frequently shown offset below the periodic table

Faraday's constant (F): charge on 1 mol of electrons; F = 96,485 C/mol e-

first transition series: transition elements in the fourth period of the periodic table (first row of the d-block), atomic numbers 21–29

formation constant (Kf): (also, stability constant) equilibrium constant for the formation of a complex ion from its components

fourth transition series: transition elements in the seventh period of the periodic table (fourth row of the d-block), atomic numbers 89 and 104–111

Frasch process: important in the mining of free sulfur from enormous underground deposits

fuel cell: devices similar to galvanic cells that require a continuous feed of redox reactants; also called a flow battery

galvanic (voltaic) cell: electrochemical cell in which a spontaneous redox reaction takes place; also called a voltaic cell

galvanization: method of protecting iron or similar metals from corrosion by coating with a thin layer of more easily oxidized zinc.

geometric isomers: isomers that differ in the way in which atoms are oriented in space relative to each other, leading to different physical and chemical properties

Gibbs free energy change (G): thermodynamic property defined in terms of system enthalpy and entropy; all spontaneous processes involve a decrease in G

group: vertical column of the periodic table

Haber process: main industrial process used to produce ammonia from nitrogen and hydrogen; involves the use of an iron catalyst and elevated temperatures and pressures

half cell: component of a cell that contains the redox conjugate pair ("couple") of a single reactant

halide: compound containing an anion of a group 17 element in the 1- oxidation state (fluoride, F-; chloride, Cl-; bromide, Br-; and iodide, I-)

Hall–Héroult cell: electrolysis apparatus used to isolate pure aluminum metal from a solution of alumina in molten cryolite

halogen: element in group 17

Henderson-Hasselbalch equation: logarithmic version of the acid ionization constant expression, conveniently formatted for calculating the pH of buffer solutions

heterogeneous catalyst: catalyst present in a different phase from the reactants, furnishing a surface at which a reaction can occur

heterogeneous equilibria: equilibria in which reactants and products occupy two or more different phases

high-spin complex: complex in which the electrons maximize the total electron spin by singly populating all of the orbitals before pairing two electrons into the lower-energy orbitals

homogeneous catalyst: catalyst present in the same phase as the reactants

homogeneous equilibria: equilibria in which all reactants and products occupy the same phase

homonuclear diatomic molecule: molecule consisting of two identical atoms

Hund's rule: every orbital in a subshell is singly occupied with one electron before any one orbital is doubly occupied, and all electrons in singly occupied orbitals have the same spin

hydrate: compound containing one or more water molecules bound within its crystals

hydrogen carbonate: salt of carbonic acid, H_2CO_3 (containing the anion HCO_3-) in which one hydrogen atom has been replaced; an acid carbonate; also known as *bicarbonate ion*

hydrogen halide: binary compound formed between hydrogen and the halogens: HF, HCl, HBr, and HI

hydrogenation: addition of hydrogen (H_2) to reduce a compound

hydrometallurgy: process in which a metal is separated from a mixture by first converting it into soluble ions, extracting the ions, and then reducing the ions to precipitate the pure metal

hydroxide: compound of a metal with the hydroxide ion OH- or the group -OH

hypothesis: tentative explanation of observations that acts as a guide for gathering and checking information

inert electrode: electrode that conducts electrons to and from the reactants in a half-cell but that is not itself oxidized or reduced

inert gas: (also, noble gas) element in group 18

inert pair effect: tendency of heavy atoms to form ions in which their valence s electrons are not lost

inner transition metal: (also, lanthanide or actinide) element in the bottom two rows; if in the first row, also called lanthanide, or if in the second row, also called actinide

interhalogen: compound formed from two or more different halogens

ion-product constant for water (Kw): equilibrium constant for the autoionization of water

ionic bond: electrostatic forces of attraction between the oppositely charged ions of an ionic compound

ionic compound: compound composed of cations and anions combined in ratios, yielding an electrically neutral substance

ionization energy: energy required to remove an electron from a gaseous atom or ion isoelectronic

ionization isomer: (or coordination isomer) isomer in which an anionic ligand is replaced by the counter ion in the inner coordination sphere

isoelectronic: group of ions or atoms that have identical electron configurations

lanthanide: inner transition metal in the top of the bottom two rows of the periodic table

lanthanide series: (also, lanthanoid series) lanthanum and the elements in the first row or the f-block, atomic numbers 57–71

lattice energy: energy requireBorn-Haber cycled to separate one mole of an ionic solid into its component gaseous ions

law: statement that summarizes a vast number of experimental observations, and describes or predicts some aspect of the natural world

law of mass action: when a reversible reaction has attained equilibrium at a given temperature, the reaction quotient remains constant

Le Châtelier's principle: an equilibrium subjected to stress will shift in a way to counter the stress and re-establish equilibrium

lead acid battery: rechargeable battery commonly used in automobiles; it typically comprises six galvanic cells based on Pb half-reactions in acidic solution

leveling effect: observation that acid-base strength of solutes in a given solvent is limited to that of the solvent's characteristic acid and base species (in water, hydronium and hydroxide ions, respectively)

Lewis acid: any species that can accept a pair of electrons and form a coordinate covalent bond

Lewis acid-base adduct: compound or ion that contains a coordinate covalent bond between a Lewis acid and a Lewis base

Lewis acid-base chemistry: reactions involving the formation of coordinate covalent bonds

Lewis base: any species that can donate a pair of electrons and form a coordinate covalent bond

ligand: ion or neutral molecule attached to the central metal ion in a coordination compound

linear: shape in which two outside groups are placed on opposite sides of a central atom

linear combination of atomic orbitals: technique for combining atomic orbitals to create molecular orbitals

linkage isomer: coordination compound that possesses a ligand that can bind to the transition metal in two different ways (CN- vs. NC-)

lithium ion battery: widely used rechargeable battery commonly used in portable electronic devices, based on lithium ion transfer between the anode and cathode

low-spin complex: complex in which the electrons minimize the total electron spin by pairing in the lower-energy orbitals before populating the higher-energy orbitals

macroscopic domain: realm of everyday things that are large enough to sense directly by human sight and touch

main-group element: (also, representative element) element in groups 1, 2, and 13–18 metal element that is shiny, malleable, good conductor of heat and electricity

metal (representative): atoms of the metallic elements of groups 1, 2, 12, 13, 14, 15, and 16, which form ionic compounds by losing electrons from their outer s or p orbitals

metalloid: element that conducts heat and electricity moderately well, and possesses some properties of metals and some properties of nonmetals noble gas (also, inert gas) element in group 18

microscopic domain: realm of things that are much too small to be sensed directly

microstate: possible configuration or arrangement of matter and energy within a system

molar solubility: solubility of a compound expressed in units of moles per liter (mol/L)

molarity (M): unit of concentration, defined as the number of moles of solute dissolved in 1 liter of solution

molecular compound: (also, covalent compound) composed of molecules formed by atoms of two or more different elements

molecular orbital: region of space in which an electron has a high probability of being found in a molecule

molecular orbital diagram: visual representation of the relative energy levels of molecular orbitals

molecular orbital theory: model that describes the behavior of electrons delocalized throughout a molecule in terms of the combination of atomic wave functions

molecular structure: structure that includes only the placement of the atoms in the molecule

monatomic ion: ion composed of a single atom

monodentate: ligand that attaches to a central metal through just one coordinate covalent bond

monoprotic acid: acid containing one ionizable hydrogen atom per molecule

Nernst equation: relating the potential of a redox system to its composition

neutral: describes a solution in which $[H3O+] = [OH-]$

nickel-cadmium battery: rechargeable battery based on Ni/Cd half-cells with applications similar to those of lithium ion batteries

nitrogen fixation: formation of nitrogen compounds from molecular nitrogen

nomenclature: system of rules for naming objects of interest

non metal: element that appears dull, poor conductor of heat and electricity period (also, series) horizontal row of the periodic table

non spontaneous process: process that requires continual input of energy from an external source

octahedral: shape in which six outside groups are placed around a central atom such that a three-dimensional shape is generated with four groups forming a square and the other two forming the apex of two pyramids, one above and one below the square plane

optical isomer: (also, enantiomer) molecule that is a nonsuperimposable mirror image with identical chemical and physical properties, except when it reacts with other optical isomers

orbital diagram: pictorial representation of the electron configuration showing each orbital as a box and each electron as an arrow

Ostwald process: industrial process used to convert ammonia into nitric acid

oxide: binary compound of oxygen with another element or group, typically containing O2- ions or the group –O– or =O

oxyacid: compound that contains hydrogen, oxygen, and one other element, bonded in a way that imparts acidic properties to the compound (ability to release H+ ions when dissolved in water)

oxyanion: polyatomic anion composed of a central atom bonded to oxygen atoms

ozone: allotrope of oxygen; O3

pairing energy (P): energy required to place two electrons with opposite spins into a single orbital

paramagnetism: phenomenon in which a material is not magnetic itself but is attracted to a magnetic field; it occurs when there are unpaired electrons present

passivation: metals with a protective nonreactive film of oxide or other compound that creates a barrier for chemical reactions; physical or chemical removal of the passivating film allows the metals to demonstrate their expected chemical reactivity

percent ionization: ratio of the concentration of ionized acid to initial acid concentration expressed as a percentage

periodic law: properties of the elements are periodic function of their atomic numbers.

periodic table: table of the elements that places elements with similar chemical properties close together

peroxide: molecule containing two oxygen atoms bonded together or as the anion, $({\text{O}}_{2}{}^{2-})$

pH: logarithmic measure of the concentration of hydronium ions in a solution

photosynthesis: process whereby light energy promotes the reaction of water and carbon dioxide to form carbohydrates and oxygen; this allows photosynthetic organisms to store energy

Pidgeon process: chemical reduction process used to produce magnesium through the thermal reaction of magnesium oxide with silicon

platinum metals: group of six transition metals consisting of ruthenium, osmium, rhodium, iridium, palladium, and platinum that tend to occur in the same minerals and demonstrate similar chemical properties

pnictogen: element in group 15

pOH: logarithmic measure of the concentration of hydroxide ions in a solution

polar covalent bond: covalent bond between atoms of different electronegativities; a covalent bond with a positive end and a negative end

polar molecule: (also, dipole) molecule with an overall dipole moment

polyatomic ion: ion composed of more than one atom

polydentate ligand: ligand that is attached to a central metal ion by bonds from two or more donor atoms, named with prefixes specifying how many donors are present (e.g., hexadentate = six coordinate bonds formed)

polymorph: variation in crystalline structure that results in different physical properties for the resulting compound

polymorph silicate: compound containing silicon-oxygen bonds, with silicate tetrahedra connected in rings, sheets, or three-dimensional networks, depending on the other elements involved in the formation of the compounds

primary cell: nonrechargeable battery, suitable for single use only

pure covalent bond: (also, nonpolar covalent bond) covalent bond between atoms of identical electronegativities

rare earth element: collection of 17 elements including the lanthanides, scandium, and yttrium that often occur together and have similar chemical properties, making separation difficult

reaction quotient (Q): mathematical function describing the relative amounts of reactants and products in a reaction mixture; may be expressed in terms of concentrations (Qc) or pressures (Qp)

representative element: (also, main-group element) element in columns 1, 2, and 12–18 series (also, period) horizontal row of the period table

representative metal: metal among the representative elements

reversible process: process that takes place so slowly as to be capable of reversing direction in response to an infinitesimally small change in conditions; hypothetical construct that can only be approximated by real processes

reversible reaction: chemical reaction that can proceed in both the forward and reverse directions under given conditions

s-p mixing: change that causes σp orbitals to be less stable than πp orbitals due to the mixing of s and p-based molecular orbitals of similar energies.

sacrificial anode: electrode constructed from an easily oxidized metal, often magnesium or zinc, used to prevent corrosion of metal objects via cathodic protection

salt bridge: tube filled with inert electrolyte solution

scientific method: path of discovery that leads from question and observation to law or hypothesis to theory, combined with experimental verification of the hypothesis and any necessary modification of the theory

second law of thermodynamics: all spontaneous processes involve an increase in the entropy of the universe

second transition series: transition elements in the fifth period of the periodic table (second row of the d-block), atomic numbers 39–47

secondary cell: battery designed to allow recharging

selective precipitation: process in which ions are separated using differences in their solubility with a given precipitating reagent

silicate: compound containing silicon-oxygen bonds, with silicate tetrahedra connected in rings, sheets, or three-dimensional networks, depending on the other elements involved in the formation of the compounds

smelting: process of extracting a pure metal from a molten ore

solubility product constant (Ksp): equilibrium constant for the dissolution of an ionic compound

solute: solution component present in a concentration less than that of the solvent

solvent: solution component present in a concentration that is higher relative to other components

spectrochemical series: ranking of ligands according to the magnitude of the crystal field splitting they induce

spontaneous change: process that takes place without a continuous input of energy from an external source

standard cell potential: the cell potential when all reactants and products are in their standard states (1 bar or 1 atm or gases; 1 M for solutes), usually at 298.15 K

standard electrode potential: electrode potential measured under standard conditions (1 bar or 1 atm for gases; 1 M for solutes) usually at 298.15 K

standard entropy (S°): entropy for one mole of a substance at 1 bar pressure; tabulated values are usually determined at 298.15 K

standard entropy change (Delta S°): change in entropy for a reaction calculated using the standard entropies

standard free energy change (ΔG°): change in free energy for a process occurring under standard conditions (1 bar pressure for gases, 1 M concentration for solutions)

standard free energy of formation (Gf): change in free energy accompanying the formation of one mole of substance from its elements in their standard states

standard hydrogen electrode (SHE): half-cell based on hydrogen ion production, assigned a potential of exactly 0 V under standard state conditions, used as the universal reference for measuring electrode potential

steel: material made from iron by removing impurities in the iron and adding substances that produce alloys with properties suitable for specific uses

stepwise ionization: process in which a polyprotic acid is ionized by losing protons sequentially

strong-field ligand: ligand that causes larger crystal field splittings

superconductor: material that conducts electricity with no resistance

symbolic domain: specialized language used to represent components of the macroscopic and microscopic domains, such as chemical symbols, chemical formulas, chemical equations, graphs, drawings, and calculations

T_2g orbitals: set of three d orbitals aligned between the Cartesian axes for coordination complexes; in octahedral complexes, they are lowered in energy compared to the eg orbitals according to CFT

tetrahedral: shape in which four outside groups are placed around a central atom such that a three-dimensional shape is generated with four corners and 109.5° angles between each pair and the central atom

theory: well-substantiated, comprehensive, testable explanation of a particular aspect of nature

third law of thermodynamics: entropy of a perfect crystal at absolute zero (0 K) is zero

third transition series: transition elements in the sixth period of the periodic table (third row of the d-block), atomic numbers 57 and 72–79

titration curve: plot of some sample property (such as pH) versus volume of added titrant

trans configuration: configuration of a geometrical isomer in which two similar groups are on opposite sides of an imaginary reference line on the molecule

transition metal: element in groups 3–12 (more strictly defined, 3–11; see chapter on transition metals and coordination chemistry)

trigonal bipyramidal: shape in which five outside groups are placed around a central atom such that three form a flat triangle with 120° angles between each pair and the central atom, and the other two form the apex of two pyramids, one above and one below the triangular plane

trigonal planar: shape in which three outside groups are placed in a flat triangle around a central atom with 120° angles between each pair and the central atom

triprotic acid: acid that contains three ionizable hydrogen atoms per molecule

valence electrons: electrons in the outermost or valence shell (highest value of n) of a ground-state atom

valence shell: outermost shell of electrons in a ground-state atom

valence shell electron-pair repulsion theory (VSEPR): theory used to predict the bond angles in a molecule based on positioning regions of high electron density as far apart as possible to minimize electrostatic repulsion

vector: quantity having magnitude and direction

weak-field ligand: ligand that causes small crystal field splittings

REFERENCES

1. Abegg R. Die Valenz Und das Periodische System. Versuch Einer Theorie der Molekularverbindungen. Z. Anorg. Chem. 1904;39:330–380. doi: 10.1002/zaac.19040390125.

2. Abegg R. Einige Bemerkungen Zur Valenztheorie. Z. Anorg. Chem. 1905;43:116–121. doi: 10.1002/zaac.19050430112.

3. Abegg R., Hinrichsen F.W. Zum Valenzbegriff. Z. Anorg. Chem. 1905;43:122–124. doi: 10.1002/zaac.19050430113.

4. Adamo C., Jacquemin D. The Calculations of Excited-State Properties with Time-Dependent Density Functional Theory. Chem. Soc. Rev. 2013;42:845–856. doi: 10.1039/C2CS35394F.

5. Arabatzis T. Representing Electrons. University of Chicago Press; Chicago, IL, USA: 2006.

6. Assis A.K.T. The Experimental and Historical Foundations of Electricity. Apeiron; Montreal, QC, Canada: 2010.

7. Bacciagaluppi G., Valentini A. Quantum Theory at the Crossroads: Reconsidering the 1927 Solvay Conference. Cambridge University Press; Cambridge, UK: 2009.

8. Bacon F. In: Novum Organum. Devey J., editor. P.F. Collier and Son; New York, NY, USA: 1902.

9. Bacon F. In: The Philosophical Works of Francis Bacon, Baron of Verulam, Viscount St. Albans, and Lord High-Chancellor of England; Methodized, and Made English from the Originals with Occasional Notes, to Explain What is Obscure; and Shew How Far the Several Plans of the Author for the Advancement of All the Parts of Knowledge, Ahve been Executed to the Present Time. 3 Volumes. Shaw P., editor. Forgotten Books; London, UK: 1783.

10. Bacon F. In: The Works of Francis Bacon. Spedding J., Ellis R.L., Heath D.D., editors. Cambridge University Press; Cambridge, UK: 2011.

11. Bader R.F.W. A Quantum Theory of Molecular Structure and Its Applications. Chem. Rev. 1991;91:893–928. doi: 10.1021/cr00005a013.

12. Bader R.F.W. The Quantum Mechanical Basis of Conceptual Chemistry. Mon. Chem. 2005;136:819–854. doi: 10.1007/s00706-005-0307-x.

13. Bader R.F.W. Atoms in Molecules. A Quantum Theory. Oxford University Press; Oxford, UK: 1994.

14. Badger G.M. Aromatic Character and Aromaticity. Cambridge University Press; London, UK: 1969.

15. Baigre B.S. Electricity and Magnetism. A Historical Perspective. Greenwood Press; Westport, CT, USA: 2007.

16. Balmer J.J. Notiz Über Die Spectrallinien des Wasserstoffs. Ann. Phys. (Berl. Ger.) 1885;261:80–87. doi: 10.1002/andp.18852610506.

17. Bardají M., Laguna A. Gold Chemistry: The Aurophilic Attraction. J. Chem. Educ. 1999;76:201. doi: 10.1021/ed076p201.

18. Batchelor J.D., Carpenter E.E., Holder G.N., Eagle C.T., Fielder J., Cummings J. Recreation of Wöhler's Synthesis of Urea: An Undergraduate Organic Laboratory Exercise. Chem. Educ. 1998;3:1–7. doi: 10.1007/s00897980271a.

19. Becke A.D. Perspective: Fifty Years of Density-Functional Theory in Chemical Physics. J. Chem. Phys. 2014;140:18A301. doi: 10.1063/1.4869598.

20. Becke A.D., Edgecombe K.E. A Simple Measure of Electron Localization in Atomic and Molecular Systems. J. Chem. Phys. 1990;92:5397–5403. doi: 10.1063/1.458517.

21. Bergman T. Dissertation on Elective Attractions. By Torbern Bergmann. Late Professor of Chemistry at Upsal, and Knight of the Royal Order of Vasa. Translated from the Latin by the Translator of Spallanzani's Dissertations. J. Murray and Charles Elliot; London, Edinburgh, UK: 1785.

22. Bergman T. Traité Des Affinités Chymiques, Ou Attractions Électives; Traduit Du Latin, Sur La Derniere Edition De Bergman. Augmenté. D'un Supplément & De Notes. Avec Des Planches. Buisson; Paris, France: 1788.

23. Bernstein J. Quantum Leaps. Belknap Press of Harvard University; Cambridge, MA, USA: 2009.

24. Berzelius J. Ansichten in Betreff der Organischen Zusammensetzung. Ann. Phys. Chem. 1846;78:161–188. doi: 10.1002/andp.18461440602.

25. Berzelius J.J. Essay on the Cause of Chemical Proportions, and on Some Circumstances Relating to Them; Together with a Short and Easy Method of Expressing Them. Ann. Philos. 1814;3:51–62.

26. Berzelius J.J. Essay on the Cause of Chemical Proportions, and on Some Circumstances Relating to Them; Together with a Short and Easy Method of Expressing Them. Ann. Philos. 1814;3:93–106.

27. Berzelius J.J. Essay on the Cause of Chemical Proportions, and on Some Circumstances Relating to Them; Together with a Short and Easy Method of Expressing Them. Ann. Philos. 1814;3:244–255.

28. Berzelius J.J. Essay on the Cause of Chemical Proportions, and on Some Circumstances Relating to Them; Together with a Short and Easy Method of Expressing Them. Ann. Philos. 1814;3:353–364.

29. Berzelius J.J. Essay on the Cause of Chemical Proportions, and on Some Circumstances Relating to Them; Together with a Short and Easy Method of Expressing Them. Ann. Philos. 1813;2:443–454.

30. Berzelius J.J. Über Die Zusammensetzung der Weinsäure Und Traubensäure (John's Säure Aus Den Vogesen), Über Das Atomengewicht Des Bleioxyds, Nebst Allgemeinen Bemerkungen Über Solche Körper, Die Gleiche Zusammensetzung, Aber Ungleiche Eigenschaften Besitzen. Ann. Phys. Phys. Chem. 1830;19:305–335. doi: 10.1002/andp.18300950702.

31. Berzelius J.J. Jahres-Bericht Über Die Fortschritte der Chemie Und Mineralogie. Laupp'sche Buchhandlung; Tübingen, Germany: 1842.

32. Bethe H. Termaufspaltung in Kristallen. Ann. Phys. (Berl. Ger.) 1929;395:133–208. doi: 10.1002/andp.19293950202.

33. Bingham R.C., Dewar M.J.S., Lo D.H. Ground States of Molecules. XXV. MINDO/3. Improved Version of the MINDO Semiempirical SCF-MO Method. J. Am. Chem. Soc. 1975;97:1285–1293. doi: 10.1021/ja00839a001.

34. Blomstrand C.W. Die Chemie der Jetztzeit Vom Standpunkte der Electrochemischen Auffassung. Aus Berzellius Lehre Entwickelt. Carl Winter's Universitätsbuchhandlung; Heidelberg, Germany: 1869.

35. Bohr N. Atomic Structure. Nature. 1921;107:104–107. doi: 10.1038/107104a0.

36. Bohr N. On the Constitution of Atoms and Molecules. Philos. Mag. 1913;26:476–502. doi: 10.1080/14786441308634993.

37. Bohr N. On the Constitution of Atoms and Molecules. Philos. Mag. 1913;26:857–875. doi: 10.1080/14786441308635031.

38. Bohr N. The Spectra of Helium and Hydrogen. Nature. 1913;92:231–232. doi: 10.1038/092231d0. [

39. Bohr N.I. On the Constitution of Atoms and Molecules. Philos. Mag. 1913;26:1–25. doi: 10.1080/14786441308634955.

40. Bolcer J.D., Hermann R.B. The Development of Computational Chemistry in the United States. In: Lipkowitz K.B., Boyd D.B., editors. Reviews in Computational Chemistry, Vol. 5. John Wiley & Sons, Inc.; Hoboken, NJ, USA: 2007. pp. 1–63.

41. Borg, R.J. and G.J. Dienes, Physical Chemistry of Solids, Academic Press, 1992.

42. Borin A.C., Gobbo J.P., Roos B.O. A Theoretical Study of the Binding and Electronic Spectrum of the Mo2 Molecule. Chem. Phys. 2008;343:210–216. doi: 10.1016/j.chemphys.2007.05.028.

43. Borin A.C., Gobbo J.P., Roos B.O. Electronic Structure and Chemical Bonding in W2 Molecule. Chem. Phys. Lett. 2010;490:24–28. doi: 10.1016/j.cplett.2010.03.022.

44. Boyle R. The Correspondence of Robert Boyle Vol. 3. Pickering & Chatto; London, UK: 2001.

45. Boyle R. The Origine of Formes and Qualities, (According to the Corpuscular Philosophy) Illustrated By Considerations and Experiments (Written Formerly By Way of Notes Upon an Essay About Nitre) By the Honourable Robert Boyle, Fellow of the Royal Society. H. Hall for Ric. Davis; Oxford, UK: 1666.

46. Boyle R. The Sceptical Chymist: Or Chymico-Physical Doubts & Paradoxes. J. Cadwell; London, UK: 1661.

47. Boyle R. The Works of Robert Boyle. Pickering & Chatto; London, UK: 1999.

48. Boys S.F. Construction of Some Molecular Orbitals to be Approximately Invariant for Changes from One Molecule to Another. Rev. Mod. Phys. 1960;32:296–299. doi: 10.1103/RevModPhys.32.296.

49. Boys S.F. Electronic Wave Functions—I. A General Method of Calculation for the Stationary States of Any Molecular System. Proc. R. Soc. Lond. Ser. A. 1950;200:542–doi: 10.1098/rspa.1950.0036.

50. Boys S.F. The Integral Formulae for the Variational Solution of the Molecular Many-Electron Wave Equation in Terms of Gaussian Functions with Direct Electronic Correlation. Proc. R. Soc. Lond. Ser. A. 1960;258:402–411. doi: 10.1098/rspa.1960.0195. [CrossRef] [Google Scholar]

51. Boys S.F., Cook G.B. Mathematical Problems in the Complete Quantum Predictions of Chemical Phenomena. Rev. Mod. Phys. 1960;32:285–295. doi: 10.1103/RevModPhys.32.285.]

52. Boys S.F., Cook G.B., Reeves C.M., Shavitt I. Automatic Fundamental Calculations of Molecular Structure. Nature (Lond. UK) 1956;178:1207–1209. doi: 10.1038/1781207a0.

53. Boys S.F., Price V.E. Electronic Wave Functions—A Calculation of Eight Variational Wave Functions for C1, C1-, S and S. Philos. Trans. R. Soc. A. 1954;246:451–462. doi: 10.1098/rsta.1954.0004.

54. Boys S.F., Sahni R.C. Electronic Wave Functions—The Evaluation of the General Vector-Coupling Coefficients by Automatic Computation. Philos. Trans. R. Soc. A. 1954;246:463–479. doi: 10.1098/rsta.1954.0005.

55. Bradley J. In: Before and After Cannizzaro: A Philosophical Commentary on the Development of the Atomic and Molecular Theories. Bradley J., editor. Whittles Publishing; North Ferriby, UK: 1992.

56. Brooke J.H. Laurent, Gerhardt, and the Philosophy of Chemistry. Hist. Stud. Phys. Sci. 1975;6:405–429. doi: 10.2307/27757345.

57. Brown A.C. On the Theory of Isomeric Compounds. J. Chem. Soc. 1865;18:230–245. doi: 10.1039/JS8651800230.

58. Brown A.C. On the Theory of Isomeric Compounds. Trans. Roy. Soc. Edin. 1864;23:707–719. doi: 10.1017/S0080456800020007.

59. Brumbaugh R.S. The Philosophers of Greece. SUNY Press; Albany, NY, USA: 1981.

60. Brundell B. Pierre Gassendi: From Aristotelianism to a New Natural Philosophy. D. Reidel; Dordrecht, Germany: 1987.

61. Brynda M., Gagliardi L., Widmark P.-O., Power P.P., Roos B.O. A Quantum Chemical Study of the Quintuple Bond Between Two Chromium Centers in [PhCRCRPh]: Trans-Bent Versus Linear Geometry. Angew. Chem. Int. Ed. 2006;45:3804–3807. doi: 10.1002/anie.200600110.

62. Buchwald J.Z., Warwick A. Histories of the Electron. MIT Press; Cambridge, MA, USA: 2004.

63. Bursten B.E., Cotton F.A., Hall M.B. Dimolybdenum: Nature of the Sextuple Bond. J. Am. Chem. Soc. 1980;102:6348–6349. doi: 10.1021/ja00540a034

64. Butlerov A. Lehrbuch Der Organischen Chemie: Zur Einführung in Das Specielle Studium Derselben. Quandt & Händel; Leipzig, Germany: 1867.

65. Cavendish H. In: The Scientific Papers of the Honourable Henry Cavendish, FRS. Thorpe E., editor. Cambridge University Press; Cambridge, UK: 1921.

66. Chachiyo T., Chachiyo H. Understanding Electron Correlation Energy Through Density Functional Theory. Comput. Chem. 2020;1172:112669. doi: 10.1016/j.comptc.2019.112669.

67. Cohen A.J., Mori-Sánchez P., Yang W. Challenges for Density Functional Theory. Chem. Rev. 2012;112:289–320. doi: 10.1021/cr200107z.

68. Cohen P.S., Cohen S.M. Wöhler's Synthesis of Urea: How Do the Textbooks Report it. J. Chem. Educ. 1996;73:883. doi: 10.1021/ed073p883.

69. Constable E.C. From Glyph to Element Symbol—A Story of Names. Chimia. 2019;73:837–839. doi: 10.2533/chimia.2019.837.

70. Constable E.C., Housecroft C.E. Coordination Chemistry: The Scientific Legacy of Alfred Werner. Chem. Soc. Rev. 2013;42:1429–1439. doi: 10.1039/C2CS35428D

71. Cook M., Karplus M. Electron Correlation and Density-Functional Methods. J. Phys. Chem. 1987;91:31–37. doi: 10.1021/j100285a010.

72. Cooke H. A Historical Study of Structures for Communication of Organic Chemistry Information Prior to 1950. Org. Biomol. Chem. 2004;2:3179–3191. doi: 10.1039/b409980j.

73. Cortés-Guzmán F., Bader R. Complementarity of QTAIM and MOTheory in the Study of Bonding in Donor-Acceptor Complexes. Coord. Chem. Rev. 2005;249:633–662. doi: 10.1016/j.ccr.2004.08.022.

74. Cotton F.A., Harris C.B. The Crystal and Molecular Structure of Dipotassium Octachlorodirhenate(III) Dihydrate, K2[Re2Cl8].2H2O. Inorg. Chem. 1965;4:330–333. doi: 10.1021/ic50025a015.

75. Cotton F.A., Murillo C.A., Walton R.A. Multiple Bonds Between Metal Atoms. Springer Science & Business Media; New York, NY, USA: 2005.

76. Coulson C.A. The Shape and Structure of Molecules. Oxford University Press; Oxford, UK: 1982.

77. Coulson C.A., O'Leary B., Mallion R.B. Hückel Theory for Organic Chemists. Academic Press; London, UK: 1978.

78. Couper A.S. On a New Chemical Theory. Philos. Mag. (1798–1977) 1858;16:104–116. doi: 10.1080/14786445808642541.

79. Couper A.S. Sur Une Nouvelle Théorie Chimique. Ann. Chim. Phys. 1858;53:469–489.

80. Couper A.S. Sur Une Nouvelle Théorie Chimique. C. R. Hebd. Seances Acad. Sci. 1858;46:1157–1160.

81. Coward H.F., Harden A. John Dalton's Lectures and Lecture Illustrations Part III. Mem. Proc. Manch. Lit. Philos. Soc. 1915;59:41–66.

82. Crosland M.P. Historical Studies in the Language of Chemistry. Dover Publ.; New York, NY, USA: 1978.

83. Dalton J. Remarks on the Essay of Dr. Berzelius on the Cause of Chemical Proportions. Ann. Philos. 1814;3:174–180.

84. Dalton J. A New System of Chemical Philosophy. Part 2. R. Bickerstaff; London, UK: 1810.

85. Dalton J. A New System of Chemical Philosophy. Part I. R. Bickerstaff; London, UK: 1808.

86. Davy H.I. The Bakerian Lecture. An Account of Some New Analytical Researches on the Nature of Certain Bodies, Particularly the Alkalies, Phosphorus, Sulphur, Carbonaceous Matter, and the Acids Hitherto Undecomposed; With Some General Observations on Chemical Theory. Philos. Mag. 1809;34:6–19. doi: 10.1080/14786440908562917.

87. De Lacy P.E. Galen on the Elements According to Hippocrates (Corpus Medicorum Graecorum) De Gruyter; Berlin, Germany: 2015.

88. Delort M. Précis De Chimie Organique (1844–1846) and the Traité De Chimie Organique (1853–1856) From Charles Gerhardt. Rev. Hist. Pharm. 2007;55:173–182. doi: 10.3406/pharm.2007.6330.

89. Descartes R. Principia Philosophiæ. Apud Ludovicum Elzevirium; Amsterdam, The Netherlands: 1644.

90. Descartes R. Principles of Philosophy. D. Reidel; Dordrecht, Germany: 1982.

91. Desiraju G., Steiner T. The Weak Hydrogen Bond. Oxford University Press; Oxford, UK: 2001.

92. Desiraju G.R. A Bond by Any Other Name. Angew. Chem. Int. Ed. Engl. 2011;50:52–59. doi: 10.1002/anie.201002960.

93. Dewar M.J.S., Thiel W. Ground States of Molecules. 38. The MNDO Method. Approximations and Parameters. J. Am. Chem. Soc. 1977;99:4899–4907. doi: 10.1021/ja00457a004.

94. Dewar M.J.S., Zoebisch E.G., Healy E.F., Stewart J.J.P. Development and Use of Quantum Mechanical Molecular Models. 76. AM1: A New General Purpose Quantum Mechanical Molecular Model. J. Am. Chem. Soc. 1985;107:3902–3909. doi: 10.1021/ja00299a024.

95. Dhanani A. "Atomism in Islamic Thought" In: Selin H., editor. Encyclopaedia of the History of Science, Technology, and Medicine in Non-Western Cultures. Springer Netherlands; Dordrecht, The Netherlands: 2016.

96. Dickerson J. Charles Gerhardt and the Theory of Organic Combination. J. Chem. Educ. 1985;62:323. doi: 10.1021/ed062p323.

97. Diels H. Die Fragmente der Vorsokratiker 1. Cambridge University Press; Cambridge, UK: 2018.

98. Diels H. Die Fragmente der Vorsokratiker 2. Inktank Publishing; Einbeck, Germany: 2019.

99. Diels H. Die Fragmente der Vorsokratiker 3. Inktank Publishing; Einbeck, Germany: 2019.

100. Diels H. Die Fragmente der Vorsokratiker 4. Inktank Publishing; Einbeck, Germany: 2019.

101. Diels H. Doxographi Graeci. Cambridge University Press; Cambridge, UK: 2010.

102. Driesch H. The History and Theory of Vitalism. Macmillan; London, UK: 1914.

103. Du Fay C.F.C. Quatrième Mémoire Sur L'Électricité, De L'Attraction & Répulsion Des Corps Electriques. [(accessed on 1 June 2020)];Mem. Acad. R. Sci. Paris. 1733 :457–476.

104. Du Fay C.F.D.C. A Letter from Mons. Du Fay, F.R.S. And of the Royal Academy of Sciences at Paris, to His Grace Charles Duke of Richmond and Lenox, Concerning Electricity. Translated from the French by T. S. M D. Philos. Trans. R. Soc. Lond. 1733;38:258–266.

105. Dunnington B.D., Schmidt J.R. Generalization of Natural Bond Orbital Analysis to Periodic Systems: Applications to Solids and Surfaces Via Plane-Wave Density Functional Theory. J. Chem. Theory. Comput. 2012;8:1902–1911. doi: 10.1021/ct300002t.

106. Eckart C. The Theory and Calculation of Screening Constants. Phys. Rev. 1930;36:878–892. doi: 10.1103/PhysRev.36.878.

107. Esteban S. Liebig–Wöhler Controversy and the Concept of Isomerism. J. Chem. Educ. 2008;85:1201. doi: 10.1021/ed085p1201.

108. Falk K.G. Some Further Considerations in the Development of the Electron Conception of Valence. Proc. Am. Philos. Soc. 1914;53:25–30.

109. Fitzgerald G.F. Helmholtz Memorial Lecture. J. Chem. Soc. Trans. 1896;69:885–912. doi: 10.1039/ct8966900885.

110. Fleming, D.G., Manz, J., Sato, K., and Takayanagi, T. (2014). Fundamental change in the nature of chemical bonding by isotopic substitution. Angewandte Chemie International Edition, 53(50): 13706–13709.

111. Fock V. "Selfconsistent Field" Mit Austausch Für Natrium. Z. Phys. 1930;62:795–805. doi: 10.1007/BF01330439.

112. Fock V. Näherungsmethode Zur Lösung des Quantenmechanischen Mehrkörperproblems. Z. Phys. 1930;61:126–148. doi: 10.1007/BF01340294.

113. Foster J.M., Boys S.F. A Quantum Variational Calculation for Hcho. Rev. Mod. Phys. 1960;32:303–304. doi: 10.1103/RevModPhys.32.303.

114. Foster J.M., Boys S.F. Quantum Variational Calculations for a Range of CH2configurations. Rev. Mod. Phys. 1960;32:305–307. doi: 10.1103/RevModPhys.32.305.

115. Fowler R.H. Bohr's Atom in Relation to the Problem of Covalency. Trans. Faraday Soc. 1923;19:459–468. doi: 10.1039/tf9231900459.

116. Frankland E. On a New Series of Organic Bodies Containing Metals. Philos. Trans. R. Soc. Lond. 1852;142:417–444. doi: 10.1098/rstl.1852.0020.

117. Franklin B. In: Letters and Papers on Electricity. Sparks J., editor. Childs and Peterson; Philadelphia, PA, USA: 1840.

118. Franklin B. New Experiments and Observations on Electricity Made at Philadelphia in America by Benjamin Franklin, Esq; and Communicated in Several Letters to Peter Collinson, Esq; of London, F.R.S. Part I. D Henry at St John's Gate; London, UK: 1760.

119. Frenking G., Shaik S. The Chemical Bond. Chemical Bonding Across the Perodic Table. Wiley-VCH; Weinheim, Germany: 2014.

120. Frenking G., Shaik S. The Chemical Bond. Fundamental Aspects of Chemical Bonding. Wiley-VCH; Weinheim, Germany: 2014.

121. Gagliardi L., Roos B.O. Quantum Chemical Calculations Show That the Uranium Molecule U2 Has a Quintuple Bond. Nature (Lond. UK) 2005;433:848–851. doi: 10.1038/nature03249.

122. Galilei G. Discorso Intorno Alle Cose Che Stanno in Su L'Acqua O Che in Quella Si Muovono. Cosimo Giunti; Florence, Italy: 1612.

123. Galilei G. Discoveries and Opinions of Galileo. 24th ed. Anchor; New York, NY, USA: 1957.

124. Gallup G.A. A Short History of Valence Bond Theory. In: Cooper D.L., editor. Valence Bond Theory. Elsevier; Amsterdam, The Netherlands: 2002. pp. 1–39.

125. Gangopadhyaya M. Indian Atomism. K.P. Balchi; Calcutta, India: 1980.

126. Gassendi P. Syntagma Philosophiae Epicuri Cum Refutationibus Dogmatum Quae Contra Fidem Christianam Ab Eo Asserta Sunt. Guillaume Barbier; Lyon, France: 1649.

127. Gay-Lussac T. Extrait De Plusieurs Notes Sur Les Métaux De La Potasse Et De La Soude, Lues À L'Institut Depuis Le 12 Janvier Jusqu'Au 16 Mai. Gaz. Natl. Ou Le Monit. Univers. 1808;40:581–582.

128. Gee W.H.H. John Dalton's Lectures and Lecture Illustrations Parts I and II. Mem. Proc. Manch. Lit. Philos. Soc. 1915;59:1–40.

129. Geoffroy E.F. Table Des Différens Rapports Observés En Chimie Entre Différentes Substances. Mém. Acad. R. Sci. 1718:202–212.

130. Geoffroy E.F. Histoire de l'Académie Royales des Sciences de Paris. Imprimerie Nationale; Paris, France: 1718. Sur Les Rapports des Différentes Substances En Chimie.

131. Gerhardt C. Précis De Chimie Organique. Fortin, Masson et Cie; Paris, France: 1845.

132. Gerhardt C. Précis De Chimie Organique. Fortin, Masson et Cie; Paris, France: 1844.

133. Gerhardt C.-F. Sur La Classification Chimique des Substances Organiques. Rev. Sci. Indust. Paris. 1843;12:592–800.

134. Gerhardt C.-F. Sur La Constitution des Sels Organiques À Acides Complexes, Et Leurs Rapports Avec Les Sels Ammoniacaux. Ann. Chim. Phys. 1839;72:184–214.

135. Gilbert W. De Magnete, Magnetisque Corporoibus, Et De Magno Magnete Tellure: Physiologia Noua, Plurimis & Argumentis, & Experimentis Demonstrata. Peter Short; London, UK: 1600.

136. Glasner R. Averroes' Physics: A Turning Point in Medieval Natural Philosophy. Oxford University Press; New York, NY, USA: 2009.

137. Glasstone S. The Structure of Some Molecular Complexes in the Liquid Phase. Trans. Faraday Soc. 1937;33:200. doi: 10.1039/tf9373300200.]

138. Gleiter R., Haberhauer G. Aromaticity and Other Conjugation Effects. Wiley-VCH; Weinheim, Germany: 2012.

139. Glendening E.D., Landis C.R., Weinhold F. Natural Bond Orbital Methods. Wires Comput. Mol. Sci. 2012;2:1–42. doi: 10.1002/wcms.51.

140. Goodgame M.M., Goddard W.A. The "sextuple" Bond of Chromium Dimer. J. Phys. Chem. 1981;85:215–217. doi: 10.1021/j150603a001.

141. Gray B.F., Pritchard H.O. A New Solution of the Wave Equation for H. J. Chem. Soc. 1957:3578. doi: 10.1039/jr9570003578.

142. Gray, H.B., Chemical Bonds, Benjamin, 1973.

143. Grisar H. Galileistudien: Historisch-Theologische Untersuchungen Über Die Urtheile Der Römischen Congregationen Im Galilei-Process. F. Pustet; Regensburg, Germany: 1882.

144. Gu Y., Kar T., Scheiner S. Fundamental Properties of the CH⋯O Interaction: Is it a True Hydrogen Bond. J. Am. Chem. Soc. 1999;121:9411–9422. doi: 10.1021/ja991795g.

145. Hall G.G. The Molecular Orbital Theory of Chemical Valency VIII. A Method of Calculating Ionization Potentials. Proc. R. Soc. Lond. Ser. A. 1951;205:541–552. doi: 10.1098/rspa.1951.0048.

146. Hammes-Schiffer S. A Conundrum for Density Functional Theory. Science. 2017;355:28–29. doi: 10.1126/science.aal3442.

147. Hartree D.R. The Wave Mechanics of an Atom with a Non-Coulomb Central Field. Part II. Some Results and Discussion. Math. Proc. Camb. Philos. Soc. 1928;24:111–132. doi: 10.1017/S0305004100011920.

148. Hattab, Helen . Basso, Sebastian. In: Lawrence Nolan, editor. The Cambridge Descartes Lexicon. Cambridge University Press; Cambridge, UK: 2015. pp. 52–53.

149. Heilbronner E., Straub W. HMO Hückel Molecular Orbitals. Springer; Berlin, Germany: 1966.

150. Hein G.E. Kekulé and the Architecture of Molecules. In: Benfey O.T., editor. Kekulé Centennial. American Chemical Society; Washington, DC, USA: 1966. pp. 1–12.

151. Heitler W. Elektronenaustausch Und Molekülbildung. [(accessed on 1 June 2020)];Nachr. Ges. Wiss. GoettingenMath. Phys. Kl. 1927 :368–374.

152. Heitler W. Zur Gruppentheorie der Homöopolaren Chemischen Bindung. Z. Phys. 1928;47:835–858. doi: 10.1007/BF01328643.

153. Heitler W. Zur Gruppentheorie der Wechselwirkung von Atomen. Z. Phys. 1928;51:805–816. doi: 10.1007/BF01400241.

154. Heitler W., Herzberg G. Eine Spektroskopische Bestätigung der Quantenmechanischen Theorie der Homopolaren Bindung. Z. Phys. 1929;53:52–56. doi: 10.1007/BF01339380.

155. Heitler W., London F. Wechselwirkung Neutraler Atome Und Homöopolare Bindung Nach der Quantenmechanik. Z. Phys. 1927;44:455–472. doi: 10.1007/BF01397394.

156. Hepler-Smith E. "Just as the Structural Formula Does": Names, Diagrams, and the Structure of Organic Chemistry at the 1892 Geneva Nomenclature Congress. Ambix. 2015;62:1–28. doi: 10.1179/1745823414Y.0000000006.

157. Higgins B. A Philosophical Essay Concerning Light. Vol. I. J. Dodsley; London, UK: 1776.]

158. Higgins W. A Comparative View of the Phlogistic and Antiphlogistic Theories. With Inductions.to Which is Annexed, an Analysis of the Human Calculus, with Observations on Its Origin, &c. 2nd ed. J. Murray; London, UK: 1791.

159. Higgins W. Experiments and Observations on the Atomic Theory, and Electrical Phenomena. Longman, Hurst, Rees, Orme and Brown; London, UK: 1814.

160. Hoffmann R. An Extended Hückel Theory. I. Hydrocarbons. J. Chem. Phys. 1963;39:1397–1412. doi: 10.1063/1.1734456.

161. Hoffmann R. Solids and Surfaces. A Chemist's View of Bonding in Extended Structures. Wiley-VCH; New York, NY, USA: 1989.

162. Hohenberg P., Kohn W. Inhomogeneous Electron Gas. Phys. Rev. 1964;136:B864–B871. doi: 10.1103/PhysRev.136.B864.

163. Holas A. Resolving Controversy About the Correlation Potential of Density-Functional Theory Far Outside a Finite System. Phys. Rev. A. 2008;78 doi: 10.1103/PhysRevA.78.014501.

164. Hückel E. Die Freien Radikale der Organischen Chemie. Z. Phys. 1933;83:632–668. doi: 10.1007/BF01330865.

165. Hückel E. Quanstentheoretische Beiträge Zum Benzolproblem. Z. Phys. 1931;72:310–337. doi: 10.1007/BF01341953.

166. Hückel E. Quantentheoretische Beiträge Zum Benzolproblem. Z. Phys. 1931;70:204–286. doi: 10.1007/BF01339530.

167. Hückel E. Quantentheoretische Beiträge Zum Problem der Aromatischen Und Ungesüttigten Verbindungen. III. Z. Phys. 1932;76:628–648. doi: 10.1007/BF01341936.

168. Hudson J. The History of Chemistry. Macmillan; Basingstoke, UK: 1992.

169. Hund F. Zur Deutung der Molekelspektren. II. Z. Phys. 1927;42:93–120. doi: 10.1007/BF01397124.

170. Hund F. Zur Deutung der Molekelspektren. IV. Z. Phys. 1928;51:759–795. doi: 10.1007/BF01400239.

171. Hund F. Linienspektren Und Periodisches System der Elemente. Springer; Vienna, Austria: 1927.

172. Hunter M. Boyle: Between God and Science. Yale University Press; New Haven, CT, USA: 2009.

173. IUPAC . Compendium of Chemical Terminology. 2nd ed. Blackwell Scientific Publications; Oxford, UK: 1997.

174. Jammer M. The Philosophy of Quantum Mechanics. The Interpretations of Quantum Mechanics in Historical Perspective. John Wiley; New York, NY, USA: 1974.

175. Jensen W.B. The Origin of the s, p, d, f Orbital Labels. J. Chem. Educ. 2007;84:757. doi: 10.1021/ed084p757.

176. Jensen W.B. The Origins of Positive and Negative in Electricity. J. Chem. Educ. 2005;82:988. doi: 10.1021/ed082p988.

177. Jörgensen S.M. Ueber Das Verhältniss Zwischen Luteo- Und Roseosalzen. J. Für Prakt. Chem. 1884;29:409–422. doi: 10.1002/prac.18840290140.

178. Joy L.S. Gassendi the Atomist. Advocate of History in an Age of Science. Cambridge University Press; Cambridge, UK: 2002.

179. Karachalios A. Erich Hückel (1896–1980) Springer; Dordrecht, Germany: 2010.

180. Kargon R.H. Atomism in England from Hariot to Newton. Cklarendon Press; Oxford, UK: 1966.

181. Kauffman G.B. Christian Wilhelm Blomstrand (1826–1897) and Sophus Mads Jørgensen (1837–1914). Their Correspondence from 1870 to 1897. Centaurus. 1977;21:44–63. doi: 10.1111/j.1600-0498.1977.tb00345.x.

182. Kauffman G.B. Sophus Mads Jorgensen (1837–1914): A Chapter in Coordination Chemistry History. J. Chem. Educ. 1959;36:521. doi: 10.1021/ed036p521.

183. Kauffman G.B. Sophus Mads Jørgensen and the Werner-Jørgensen Controversy. Chymia. 1960;6:180–204. doi: 10.2307/27757198.

184. Keith A.B. Indian Logic and Atomism; an Exposition of the Nyãya and Vaicesika Systems. Clarendon Press; Oxford, UK: 1921.

185. Kekulé A. Note Sur Quelques Produits De Substitution De La Benzine. Bull. Acad. R. Belg. 1865;2:551–563.

186. Kekulé A. Sur L'Atomicité Des Éléments. C. R. Hebd. Seances Acad. Sci. 1864;58:510–514.

187. Kekulé A. Sur La Constitution des Substances Aromatiques. Bull. De La Soc. Chim. De Fr. 1865;3:98–110.

188. Kekulé A. Über Die Constitution Und Die Metamorphosen der Chemischen Verbindungen Und Über Die Chemische Natur des Kohlenstoffs. Justus Liebigs Ann. Chem. 1858;106:129–159. doi: 10.1002/jlac.18581060202.

189. Kekulé A., Über Die S.G. Gepaarten Verbindungen Und Die Theorie der Mehratomigen Radicale. Justus Liebigs Ann. Chem. 1857;104:129–150. doi: 10.1002/jlac.18571040202.

190. Keyser P.T. The Purpose of the Parthian Galvanic Cells: A First-Century a. D. Electric Battery Used for Analgesia. J. Near East. Stud. 1993;52:81–98. doi: 10.1086/373610.

191. Kim M.G. Affinity, That Elusive Dream. A Genealogy of the Chemical Revolution. MIT; Cambridge, MA, USA: 2008.

192. King W.J. Contributions from the Museum of History and Technology. The Natural Philosophy of William Gilbert and His Predecessors. Volume 8. Bulletin of the United States National Museum; Washington, DC, USA: 1959. pp. 121–139.

193. Kinnersley E. New Experiments in Electricity: In a Letter from Mr. Ebenezer Kinnersley, to Benjamin Franklin, Ll. D. F. R. S. Philos. Trans. R. Soc. Lond. 1763;53:84–97. doi: 10.1098/rstl.1763.0022.

194. Kinne-Saffran E., Kinne R.K. Vitalism and Synthesis of Urea. From Friedrich Wöhler to Hans a. Krebs. Am. J. Nephrol. 1999;19:290–294. doi: 10.1159/000013463.

195. Kirthisinghe B.P. Buddhism and Science. Motilal Banarsidass; Delhi, India: 1984.

196. Kohn W., Sham L.J. Self-Consistent Equations Including Exchange and Correlation Effects. Phys. Rev. 1965;140:A1133–A1138. doi: 10.1103/PhysRev.140.A1133.

197. Kragh H.S.M. Jørgensen and His Controversy with a. Werner: A Reconsideration. Br. J. Hist. Sci. 1997;30:203–219. doi: 10.1017/S0007087497003014.

198. Kraus D., Lorenz M., Bondybey V.E. On the Dimers of the VIb Group: A New NIR Electronic State of Mo2. Phys. Chem. Comm. 2001;4:44. doi: 10.1039/b104063b.

199. Kreisel K.A., Yap G.P.A., Dmitrenko O., Landis C.R., Theopold K.H. The Shortest Metal-Metal Bond Yet: Molecular and Electronic Structure of a Dinuclear Chromium Diazadiene Complex. J. Am. Chem. Soc. 2007;129:14162–14163. doi: 10.1021/ja076356t.

200. Kryachko E.S., Ludeña E.V. Density Functional Theory: Foundations Reviewed. Phys. Rep. 2014;544:123–239. doi: 10.1016/j.physrep.2014.06.002.

201. Kuhn T.S. Robert Boyle and Structural Chemistry in the Seventeenth Century. Isis. 1952;43:12–36. doi: 10.1086/349360.

202. Lactantius . Divine Institutes, Books 1-7. Catholic University of America Press; Washington, DC, USA: 1964.

203. Lakhtakia A. Models and Modelers of Hydrogen. World Scientific; Singapore: 1996. [Google Scholar]

204. Laming R. Observations on a Paper by Prof. Faraday Concerning Electric Conduction and the Nature of Matter. Philos. Mag. (1798–1977) 1845;27:420–423. doi: 10.1080/14786444508646245.

205. Laming R. On the Primary Forces of Electricity. Part 1 cont. Philos. Mag. (1798–1977) 1838;13:44–54.

206. Laming R. On the Primary Forces of Electricity. Part I. Philos. Mag. (1798–1977) 1838;12:486–498.

207. Laming R. On the Primary Forces of Electricity. Part II. Philos. Mag. (1798–1977) 1838;13:336–339.

208. Langmuir, I. (1919). The arrangement of electrons in atoms and molecules. *Journal of the American Chemical Society, 41*(6): 868-934.

209. Le Bel J.-A. Sur Les Relations Qui Existent Entre Les Formules Atomiques des Corps Organiques Et Le Pouvoir Rotatoire De Leurs Dissolutions. Bull. De La Soc. Chim. De Fr. 1874;22:337–347.

210. Le Grand H.E. Galileo's Matter Theory. In: Butts R.E., Pitt J.C., editors. New Perspectives on Galileo Vol. 14. Springer; Dordrecht, The Netherlands: 1978. pp. 197–208.

211. Lemay J.A.L.L. Ebenezer Kinnersley. Franklin's Friend. University of Pennsylvania Press; Philadelphia, PA, USA: 1964.

212. Lennard-Jones J.E. The Electronic Structure of Some Diatomic Molecules. Trans. Faraday Soc. 1929;25:668. doi: 10.1039/tf9292500668.

213. Lennard-Jones J.E. The Molecular Orbital Theory of Chemical Valency.2. Equivalent Orbitals in Molecules of Known Symmetry. Proc. R. Soc. Lond. Ser. A. 1949;198:14–26. doi: 10.1098/rspa.1949.0084.

214. Lennerd-Jones J. The Molecular Orbital Theory of Chemical Valency.1. The Determination of Molecular Orbitals. Proc. R. Soc. Lond. Ser. A. 1949;198:1–13. doi: 10.1098/rspa.1949.0083.

215. Leucippus, Democritus . The Atomists, Leucippus and Democritus. Fragments: A Text and Translation with a Commentary. University of Toronto Press; Toronto, ON, Canada: 1999.

216. Levere T.H. Affinity and Matter. Elements of Chemical Philosophy. Clarendon Press; Oxford, UK: 1971. pp. 1800–1865.

217. Levere T.H. Chemists and Chemistry in Nature and Society. Routledge; Aldershot, UK: 1994. pp. 1770–1878.

218. Lewis D.E. 150 Years of Organic Structures. In: Giunta C.J., editor. Atoms in Chemistry: From Dalton's Predecessors to Complex Atoms and Beyond. American Chemical Society; Washington, DC, USA: 2010. pp. 35–37.

219. Lewis D.E. Introduction to an English Translation, "on the Different Explanations of Certain Cases of Isomerism" By Aleksandr Butlerov. Bull. Hist. Chem. 2015;40:9–12.

220. Lewis, G.N. (1916). The atom and the molecule. *Journal of the American Chemical Society, 38*(4): 762-786.

221. Liebig J. Über Wöhler's Cyansäure. Arch. Ges. Nat. 1825;6:145–153.

222. Liebig J. Ueber Cyan- Und Knallsäure. J. Chem. U. Phys. 1826;18:376–381.

223. Liebig J., Gay-Lussac J.L. Analyse Du Fulminate D'argent. Ann. Chim. Phys. 1824;25:285–311.

224. Liu W., Xiao Y. Relativistic Time-Dependent Density Functional Theories. Chem. Soc. Rev. 2018;47:4481–4509. doi: 10.1039/C8CS00175H.

225. Lodge O.J. On Nodes and Loops in Connexion with Chemical Formulae. Philos. Mag. 1875;50:367–376. doi: 10.1080/14786447508641304.

226. Löwdin P.-O. Quantum Theory of Many-Particle Systems. I. Physical Interpretations By Means of Density Matrices, Natural Spin-Orbitals, and Convergence Problems in the Method of Configurational Interaction. Phys. Rev. 1955;97:1474–1489. doi: 10.1103/physrev.97.1474.

227. Lucretius . In: De rerum natura. Deufert M., editor. Walter de Gruyter GmbH & Co KG; Berlin, Germany: 2019.

228. Lüthy C.H., Murdoch J.E., Newman W.R. Late Medieval and Early Modern Corpuscular Matter Theories. Brill; Leiden, The Netherlands: 2001.

229. Madan H.G. Remarks on Some Points in the Nomenclature of Salts. J. Chem. Soc. 1870;23:22–28. doi: 10.1039/JS8702300022.

230. Magnasco V. Models for Bonding in Chemistry. Wiley; Chichester, UK: 2010.

231. Magnus A. De Mineralibus. Johannes et Gregorius de Gregoriis; Venice, Italy: 1495.

232. Magnus A. The Book of Minerals. Clarendon Press; Oxford, UK: 1967.

233. Mardirossian N., Head-Gordon M. Thirty Years of Density Functional Theory in Computational Chemistry: An Overview and Extensive Assessment of 200 Density Functionals. Mol. Phys. 2017;115:2315–2372. doi: 10.1080/00268976.2017.1333644.

234. Marques M.A.L., Maitra N.T., Nogueira F.M.S., Gross E.K.U., Rubio A. Fundamentals of Time-Dependent Density Functional Theory. Springer-Verlag; Berlin, Germany: 2012.

235. Mason H.S. History of the Use of Graphic Formulas in Organic Chemistry. Isis. 1943;34:346–354. doi: 10.1086/347834.

236. McEvilley T. The Shape of Ancient Thought: Comparative Studies in Greek and Indian Philosophies. Constable & Robinson; New York, NY, USA: 2012.

237. Mehra J., Rechenberg H. Erwin Schrödinger and the Rise of Wave Mechanics. Part 1 Schrödinger in Vienna and Zurich 1887–1925. Springer-Verlag; New York, NY, USA: 1987.

238. Mehra J., Rechenberg H. The Historical Development of Quantum Theory, 6 Volumes. Springer-Verlag; New York, NY, USA: 1987.

239. Mehra J., Rechenberg H. The Quantum Theory of Planck, Einstein, Bohr, and Sommerfeld. Its Foundation and the Rise of Its Difficulties, 1900–1925. Springer-Verlag; New York, NY, USA: 1982.

240. Merino G., Donald K.J., D'Acchioli J.S., Hoffmann R. The Many Ways to Have a Quintuple Bond. J. Am. Chem. Soc. 2007;129:15295–15302. doi: 10.1021/ja075454b.

241. Meyer H.W. A History of Electricity and Magnetism. M.I.T. Press; Cambridge, MA, USA: 1971.

242. Meyer L. Die Modernen Theorien der Chemie Und Ihre Bedeutung Für Die Chemische Statik. von Maruschke & Berendt; Breslau, Germany: 1864.

243. Meyer L. Die Modernen Theorien der Chemie Und Ihre Bedeutung Für Die Chemische Statik. Zweite Umgearbeitete Und Sehr Vermehrte Auflag. von Maruschke & Berendt; Breslau, Germany: 1872.

244. Meyer V. Zur Valenz Und Verbindungsfähigkeit des Kohlenstoffs. Justus Liebigs Ann. Chem. 1876;180:192–206. doi: 10.1002/jlac.18761800119.

245. Michael E. Daniel Sennert on Matter and Form: At the Juncture of the Old and the New1. Early Sci. Med. 1997;2:272–299. doi: 10.1163/157338297X00159.

246. Mingos D.M.P. The Chemical Bond I. Springer International; Cham, Switzerland: 2016.

247. Mingos D.M.P. The Chemical Bond II. Springer International; Cham, Switzerland: 2016.

248. Mingos D.M.P. The Chemical Bond III. Springer International; Cham, Switzerland: 2016.

249. Minkin V.I. Current Trends in the Development of a. M. Butlerov's Theory of Chemical Structure. Russ. Chem. Bull. 2012;61:1265–1290. doi: 10.1007/s11172-012-0174-7.

250. Minkin V.I., Glukhovtsev M.N., Simkin B.Y. Aromaticity and Antiaromaticity. Wiley-Interscience; New York, NY, USA: 1994.

251. More L.T. Boyle as Alchemist. J. Hist. Ideas. 1941;2:61. doi: 10.2307/2707281.

252. Mottelay P.F. Bibliographical History of Electricity and Magnetism. Griffin; London, UK: 1922.

253. Mulliken R.S. Bonding Power of Electrons and Theory of Valence. Chem. Rev. 1931;9:347–388. doi: 10.1021/cr60034a001.

254. Mulliken R.S. Electronic States and Band Spectrum Structure in Diatomic Molecules. I. Statement of the Postulates. Interpretation of CuH, CH, and CO Band-Types. Phys. Rev. 1926;28:481–506. doi: 10.1103/PhysRev.28.481.

255. Mulliken R.S. Electronic States and Band Spectrum Structure in Diatomic Molecules. IV. Hund's Theory; Second Positive Nitrogen and Swan Bands; Alternating Intensities. Phys. Rev. 1927;29:637–649. doi: 10.1103/PhysRev.29.637.

256. Mulliken R.S. Electronic Structures of Polyatomic Molecules and Valence. Phys. Rev. 1932;40:55–62. doi: 10.1103/PhysRev.40.55.

257. Mulliken R.S. Electronic Structures of Polyatomic Molecules and Valence. II. General Considerations. Phys. Rev. 1932;41:49–71. doi: 10.1103/PhysRev.41.49.

258. Mulliken R.S. The Assignment of Quantum Numbers for Electrons in Molecules. III. Diatomic Hydrides. Phys. Rev. 1929;33:730–747. doi: 10.1103/PhysRev.33.730.

259. Mulliken R.S. The Assignment of Quantum Numbers for Electrons in Molecules. I. Phys. Rev. 1928;32:186–222. doi: 10.1103/PhysRev.32.186.

260. Mulliken R.S. The Interpretation of Band Spectra Part III. Electron Quantum Numbers and States of Molecules and Their Atoms. Rev. Mod. Phys. 1932;4:1–86. doi: 10.1103/RevModPhys.4.1.

261. Murrell J.N. The Origins and Later Developments of Molecular Orbital Theory. Int. J. Quantum Chem. 2012;112:2875–2879. doi: 10.1002/qua.23293.

262. Needham J. Science and Civilisation in China: Volume 7, the Social Background, Part 2, General Conclusions and Reflections. Cambridge University Press; Cambridge, UK: 2004.

263. Needham P. Hydrogen Bonding: Homing in on a Tricky Chemical Concept. Stud. Hist. Philos. Sci. A. 2013;44:51–65. doi: 10.1016/j.shpsa.2012.04.001.

264. Newman W. The Significance of "Chymical Atomism" Early Sci. Med. 2009;14:248–264. doi: 10.1163/157338209X425579.

265. Newman W.R. Robert Boyle, Transmutation, and the History of Chemistry Before Lavoisier: A Response to Kuhn. Osiris. 2014;29:63–77. doi: 10.1086/678097.

266. Newman W.R. The Alchemical Sources of Robert Boyle's Corpuscular Philosophy. Ann. Sci. 1996;53:567–585. doi: 10.1080/00033799600200401.

267. Newman W.R. Atoms and Alchemy: Chymistry and the Experimental Origins of the Scientific Revolution. University of Chicago Press; Chicago, IL, USA: 2006.

268. Newton I. Opticks: Or a Treatise of the Reflections, Refractions, Inflections and Colours of Light. The Fourth Edition, Corrected. William Innys; London, UK: 1730.

269. Newton, I. (1704). *Opticks: or, a treatise of the reflexions, refractions, inflexions and colours of light.*

270. Nguyen T., Sutton A.D., Brynda M., Fettinger J.C., Long G.J., Power P.P. Synthesis of a Stable Compound with Fivefold Bonding Between Two Chromium(I) Centers. Science. 2005;310:844–847. doi: 10.1126/science.1116789.

271. Normandin S., Wolfe C.T. Vitalism and the Scientific Image in Post-Enlightenment Life Science. Springer; Dordrecht, The Netherlands: 2013. pp. 1800–2010.

272. Numbers R.L., Kampourakis K. Newton's Apple and Other Myths about Science. Harvard University Press; Cambridge, MA, USA: 2015.

273. O'Keefe T. Epicureanism. Acumen Publishing; Durham, UK: 2010.

274. Otten B.M., Melançon K.M., Omary M.A. All That Glitters is Not Gold: A Computational Study of Covalent vs. Metallophilic Bonding in Bimetallic Complexes of d10 Metal Centers—A Tribute to Al Cotton on the Tenth Anniversary of His Passing. Comments Inorg. Chem. 2018;38:1–35. doi: 10.1080/02603594.2018.1467315.

275. Palmer W.G. A History of the Concept of Valency to 1930. Cambridge University Press; Cambridge, UK: 2010.

276. Paparazzo E. Vacuum: A Void Full of Questions. Surf. Interface Anal. 2008;40:450–453. doi: 10.1002/sia.2608.

277. Parr R.G., Yang W. Density-Functional Theory of Atoms and Molecules. Oxford University Press; New York, NY, USA: 1994.

278. Partington J.R. Albertus Magnus on Alchemy. Ambix. 1937;1:3–20. doi: 10.1179/amb.1937.1.1.3.

279. Partington J.R. A History of Chemistry. Vol. 1. Theoretical Background. Macmillan and Co.; London, UK: 1970.

280. Partington J.R. A History of Chemistry. Vol. 2. Macmillan and Co.; London, UK: 1961.

281. Partington J.R. A History of Chemistry. Vol. 3. Macmillan and Co.; London, UK: 1962.

282. Partington J.R. A History of Chemistry. Vol. 4. Macmillan and Co.; London, UK: 1964.

283. Partington J.R. A Short History of Chemistry. 3rd ed. Dover; New York, NY, USA: 1989.

284. Pauling L. Die Abschirmungskonstanten der Relativistischen Oder Magnetischen Röntgenstrahlendubletts. Z. Phys. 1927;40:344–350. doi: 10.1007/BF01486079.

285. Pauling L. The Nature of the Chemical Bond. Application of Results Obtained from the Quantum Mechanics and from a Theory of Paramagnetic Susceptibility to the Structure of Molecules. J. Am. Chem. Soc. 1931;53:1367–1400. doi: 10.1021/ja01355a027.

286. Pauling L. The Nature of the Chemical Bond. IV The Energy of Single Bonds and the Relative Electronegativity of Atoms. J. Am. Chem. Soc. 1932;54:3570–3582. doi: 10.1021/ja01348a011.

287. Pauling L. The Shared-Electron Chemical Bond. Proc. Natl. Acad. Sci. USA. 1928;14:359–362. doi: 10.1073/pnas.14.4.359.

288. Pauling L. The Nature of the Chemical Bond and the Structure of Molecules and Crystals: An Introduction to Modern Structural Chemistry. Cornell University Press; Ithaca, NY, USA: 1939.

289. Pauling L. The Nature of the Chemical Bond and the Structure of Molecules and Crystals: An Introduction to Modern Structural Chemistry. 2nd ed. Cornell University Press; Ithaca, NY, USA: 1940.

290. Pauling L. The Nature of the Chemical Bond and the Structure of Molecules and Crystals: An Introduction to Modern Structural Chemistry. 3rd ed. Cornell University Press; Ithaca, NY, USA: 1960.

291. Pauling, L. (1931). The nature of the chemical bond. Application of results obtained from the quantum mechanics and from a theory of paramagnetic susceptibility to the structure of molecules. *Journal of the American Chemical Society, 53*(4): 1367-1400.

292. Pauling, L., General Chemistry, W.H. Freeman and Co., San Francisco, 1970, Chapter 6.

293. Popelier P.L.A. The Chemical Bond. Wiley-VCH; Weinheim, Germany: 2014. The Qtaim Perspective of Chemical Bonding; pp. 271–308.

294. Pople J.A., Beveridge D.L., Dobosh P.A. Approximate Self-consistent Molecular-Orbital Theory. V. Intermediate Neglect of Differential Overlap. J. Chem. Phys. 1967;47:2026–2033. doi: 10.1063/1.1712233.

295. Pople J.A., Santry D.P., Segal G.A. Approximate Self-consistent Molecular Orbital Theory. I. Invariant Procedures. J. Chem. Phys. 1965;43:S129–S135. doi: 10.1063/1.1701475.

296. Pople J.A., Segal G.A. Approximate Self-consistent Molecular Orbital Theory. II. Calculations with Complete Neglect of Differential Overlap. J. Chem. Phys. 1965;43:S136–S151. doi: 10.1063/1.1701476.

297. Pople J.A., Segal G.A. Approximate Self-consistent Molecular Orbital Theory. III. CNDO Results for AB2 and AB3 Systems. J. Chem. Phys. 1966;44:3289–3296. doi: 10.1063/1.1727227.

298. Priestley J. In: The History and Present State of Electricity with Original Experiments. 5th Edition, Corrected. Johnson J., Rivington F.C., Cadell T., Baldwin R., Lowndes H., editors. Gale Ecco; London, UK: 1794.

299. Principe L. The Aspiring Adept. Robert Boyle and His Alchemical Quest. Princeton University Press; Princeton, NJ, USA: 2000.

300. Pritchard H.O., Sumner F.H. A Property of 'Repeating' Secular Determinants. Philos. Mag. (1798–1977) 1954;45:466–470. doi: 10.1080/14786440508521094.

301. Pritchard H.O., Sumner F.H. The Application of Electronic Digital Computers to Molecular Orbital Problems—Ii. A New Approximation for Hetero-Atom Systems. Proc. R. Soc. Lond. Ser. A. 1956;235:136–143. doi: 10.1098/rspa.1956.0070.

302. Pritchard H.O., Sumner F.H. The Application of Electronic Digital Computers to Molecular Orbital Problems I. The Calculation of Bond Lengths in Aromatic Hydrocarbons. Proc. R. Soc. Lond. Ser. A. 1954;226:128–140. doi: 10.1098/rspa.1954.0243.

303. Priyadarshi P. Zero is Not the Only Story. India First Foundation; New Delhi, India: 2007.

304. Proust J.L. D'un Mémoire Intitulé: Recherches Sur Le Bleu De Prusse. Ann. Chim. (CachanFr.) 1797;23:85–101.

305. Pullman B. The Atom in the History of Human Thought. Oxford University Press; New York, NY, USA: 1998.

306. Purrington R.D. The Heroic Age. The Creation of Quantum Mechanics, 1925–1940. Oxford University Press; Oxford, UK: 2018.

307. Pyykkö P., Li J., Runeberg N. Predicted Ligand Dependence of the Au(I)... Au(I) Attraction in (XAuPh3)2. Chem. Phys. Lett. 1994;218:133–138. doi: 10.1016/0009-2614(93)E1447-O.

308. Radhakrishnan S. A Source Book in Indian Philosophy. Princeton University Press; Princeton, NJ, USA: 1957.

309. Ramberg P.J. Johannes Wislicenus, Atomism, and the Philosophy of Chemistry. Bull. Hist. Chem. 1994;15/16:45.

310. Rees G. Atomism and 'Subtlety' in Francis Bacon's Philosophy. Ann. Sci. 1980;37:549–571. doi: 10.1080/00033798000200391.

311. Richardson G.M., Pasteur L., van 't Hoff J.H., Le Bel J.-A., Wislicenus J. The Foundations of Stereochemistry; Memoirs by Pasteur, Van't Hoff, Lebel and Wislicenus. American Book Co.; New York, NY, USA: 1901.

312. Rocke A.J. Kekulé, Butlerov, and the Historiography of the Theory of Chemical Structure. Br. J. Hist. Sci. 1981;14:27–57. doi: 10.1017/S0007087400018276.

313. Rocke A.J. The Reception of Chemical Atomism in Germany. Isis. 1979;70:519–536. doi: 10.1086/352339.

314. Scherbaum F., Grohmann A., Huber B., Krüger C., Schmidbaur H. "Aurophilicity" as a Consequence of Relativistic Effects: The Hexakis(triphenylphosphaneaurio) methane Dication. Angew. Chem. Int. Ed. Engl. 1988;27:1544–1546. doi: 10.1002/anie.198815441.

315. Schmid E., Ziegelmann H. The Quantum Mechanical Three-Body Problem. Pergamon; Oxford, UK: 1974.

316. Schmidbaur H. The Aurophilicity Phenomenon: A Decade of Experimental Findings, Theoretical Concepts and Emerging Applications. Gold Bull. (Berl. Ger.) 2000;33:3–10. doi: 10.1007/BF03215477.

317. Schmidbaur H., Schier A. A Briefing on Aurophilicity. Chem. Soc. Rev. 2008;37:1931–1951. doi: 10.1039/b708845k.

318. Schmidbaur H., Schier A. Aurophilic Interactions as a Subject of Current Research: An Up-Date. Chem. Soc. Rev. 2012;41:370–412. doi: 10.1039/C1CS15182G.

319. Schrödinger E. Bohrs Neue Strahlungshypothese Und der Energiesatz. Naturwissenschaften. 1924;12:720–724. doi: 10.1007/BF01504820.

320. Sennert D. Hypomnemata Physica. Clementis Schleichii et consortum; Frankfurt, Germany: 1636.

321. Slater J.C. Atomic Shielding Constants. Phys. Rev. 1930;36:57–64. doi: 10.1103/PhysRev.36.57.

322. Slater J.C. Molecular Orbital and Heitler–London Methods. J. Chem. Phys. 1965;43:S11–S17. doi: 10.1063/1.1701472.

323. Slater J.C. Note on Hartree's Method. Phys. Rev. 1930;35:210–211. doi: 10.1103/PhysRev.35.210.2.

324. Slater J.C. Quantum Theory of Molecules and Solids: Electronic Structure of Molecules. McGraw-Hill; New York, NY, USA: 1963.

325. Sloggett H. On the Constitution of Matter. Philos. Mag. (1798–1977) 1845;27:443–448.

326. Smith S.J., Sutcliffe B.T. The Development of Computational Chemistry in the United Kingdom. In: Lipkowitz K.B., Boyd D.B., editors. Reviews in Computational Chemistry, Vol. 10. John Wiley & Sons, Inc.; Hoboken, NJ, USA: 1996. pp. 271–313.

327. Sommerfeld A. Atomic Structure and Spectral Lines. Methuen; London, UK: 1923.

328. Stcherbatsky T. Buddhist Logic. 2vols. Dover; New York, NY, USA: 1930.

329. Steiner T., Desiraju G.R. Distinction Between the Weak Hydrogen Bond and the Van der Waals Interaction. Chem. Commun. (Camb. UK) 1998:891–892. doi: 10.1039/a708099i.

330. Stewart J.J. Optimization of Parameters for Semiempirical Methods IV: Extension of MNDO, AM1, and PM3 to More Main Group Elements. J. Mol. Model. 2004;10:155–164. doi: 10.1007/s00894-004-0183-z.

331. Stewart J.J.P. Optimization of Parameters for Semiempirical Methods I. Method. J. Comp. Chem. 1989;10:209–220. doi: 10.1002/jcc.540100208.

332. Stewart J.J.P. Optimization of Parameters for Semiempirical Methods II. Applications. J. Comp. Chem. 1989;10:221–264. doi: 10.1002/jcc.540100209.

333. Stewart J.J.P. Optimization of Parameters for Semiempirical Methods. III Extension of PM3 to Be, Mg, Zn, Ga, Ge, as, Se, Cd, In, Sn, Sb, Te, Hg, Tl, Pb, and Bi. J. Comp. Chem. 1991;12:320–341. doi: 10.1002/jcc.540120306.

334. Stolleis M. Natural Law and Laws of Nature in Early Modern Europe: Jurisprudence, Theology, Moral and Natural Philosophy. Routledge; Abingdon-on-Thames, UK: 2016.

335. Stone A.D. Einstein and the Quantum. The Quest of the Valiant Swabian. Princeton University Press; Princeton, NJ, USA: 2013.

336. Stone A.J. Natural Bond Orbitals and the Nature of the Hydrogen Bond. J. Phys. Chem. A. 2017;121:1531–1534. doi: 10.1021/acs.jpca.6b12930.

337. Tóth Z. A Demonstration of Wöhler's Experiment: Preparation of Urea from Ammonium Chloride and Potassium Cyanate. J. Chem. Educ. 1996;73:539. doi: 10.1021/ed073p539.2.

338. Van 't Hoff J.H. Sur Les Formules De Structure Dans L'Espace. Arch. Neerl. Sci. Exactes Nat. 1874;9:445–454.

339. Van 't Hoff J.H. Sur Les Formules De Structure Dans L'Espace. Bull. De La Soc. Chim. De Fr. 1875;23:295–301.

340. Van Melsen A.G. From Atomos to Atom: The History and Concept of the Atom. Dover Phoenix Editions; Mineola, NY, USA: 1952.

341. Van Vleck J.H. On the Theory of the Structure of CH4and Related Molecules. Part II. J. Chem. Phys. 1933;1:177–182. doi: 10.1063/1.1749270.

342. Van Vleck J.H. Theory of Electric and Magnetic Susceptibilities. Oxford University Press; Oxford, UK: 1932.

343. Verma P., Truhlar D.G. Status and Challenges of Density Functional Theory. Trends Chem. 2020;2:302–318. doi: 10.1016/j.trechm.2020.02.005.

344. Wambaugh J. The Delta Star. Bantam; New York, NY, USA: 1984.

345. Warren J. The Cambridge Companion to Epicureanism. Cambridge University Press; Cambridge, UK: 2009.

346. Weinhold F., Glendening E.D. Comment on "natural Bond Orbitals and the Nature of the Hydrogen Bond" J. Phys. Chem. A. 2018;122:724–732. doi: 10.1021/acs.jpca.7b08165.

347. Weinhold F., Landis C.R. Valency and Bonding. Cambridge University Press; Cambridge, UK: 2003.

348. Werner A. Beitrag Zur Konstitution Anorganischer Verbindungen. Z. Anorg. Chem. 1893;3:267–330. doi: 10.1002/zaac.18930030136.

349. Werner A. Über Haupt- Und Nebenvalenzen Und Die Constitution der Ammoniumverbindungen. Justus Liebig's Ann. Chem. 1902;322:261–296. doi: 10.1002/jlac.19023220302.

350. Werner A.S.M. Jörgensen 4. Juli 1837–1. April 1914. Chem. Ztg. 1914 Apr 19;38:557–564.

351. West, A.R., Solid State Chemistry, J. Wiley, 1984.

352. Whisnant D.M. Bonding Theory/the Werner-Jorgensen Controversy. J. Chem. Educ. 1993;70:902. doi: 10.1021/ed070p902.1.

353. Wichelhaus H. Über Die Verbindungen des Phosphors. Ann. Chem. Pharm. 1868;6:257–280.

354. Williams E.A. A Cultural History of Medical Vitalism in Enlightenment Montpellier. Ashgate Publishing, Ltd.; Burlington, VT, USA: 2003.

355. Wilson C. Epicureanism. Oxford University Press; Oxford, UK: 2015.

356. Wislicenus J. Ueber Die Räumliche Anordnung der Atome in Organischen Molekulen and Ihre Bestimmung in Geometrisch-Isomeren Ungesättigen Verbindungen. Abh. Math. Phys. Kl. K. Saechs. Ges. Wiss. 1887;14:1–77.

357. Wisniak J. Charles Fréderic Gerhardt. Educ. Quim. 2018;17:343. doi: 10.22201/fq.18708404e.2006.3.66037.

358. Wöhler F. Analytische Versuche Über Die Cyansäure. Ann. Phys. (Berl. Ger.) 1824;77:117–124. doi: 10.1002/andp.18240770506.

359. Wöhler F. Über Die Zusammensetzung der Cyansäure. Ann. Phys. (Berl. Ger.) 1825;81:385–388. doi: 10.1002/ardp.18270210118.

360. Wöhler F. Über Künstliche Bildung des Harnstoffs. Ann. Phys. (Berl. Ger.) 1828;88:253–256. doi: 10.1002/andp.18280880206.

361. Yates K. Hückel Molecular Orbital Theory. Academic Press; New York, NY, USA: 1978.

362. Zener C. Analytic Atomic Wave Functions. Phys. Rev. 1930;36:51–56. doi: 10.1103/PhysRev.36.51.

363. Zubarev D.Y., Boldyrev A.I. Deciphering Chemical Bonding in Golden Cages. J. Phys. Chem. A. 2009;113:866–868. doi: 10.1021/jp808103t.

364. Zubarev D.Y., Boldyrev A.I. Developing Paradigms of Chemical Bonding: Adaptive Natural Density Partitioning. Phys. Chem. Chem. Phys. 2008;10:5207. doi: 10.1039/b804083d.

365. Zubarev D.Y., Boldyrev A.I. Revealing Intuitively Assessable Chemical Bonding Patterns in Organic Aromatic Molecules Via Adaptive Natural Density Partitioning. J. Org. Chem. 2008;73:9251–9258. doi: 10.1021/jo801407e.

INDEX

www.ingramcontent.com/pod-product-compliance
Lightning Source LLC
Chambersburg PA
CBHW061933190326
41458CB00009B/2725